Straw Man Arguments

Also available from Bloomsbury

A Critical Introduction to Knowledge-How, by J. Adam Carter and Ted Poston
Evidentialism and the Will to Believe, by Scott Aikin
Logic of the Digital, by Aden Evens
The History of Philosophical and Formal Logic, edited by Alex Malpass
and Marianna Antonutti Marfori
Using Questions to Think, by Nathan Eric Dickman

Straw Man Arguments

A Study in Fallacy Theory

Scott Aikin and John Casey

BLOOMSBURY ACADEMIC
LONDON • NEW YORK • OXFORD • NEW DELHI • SYDNEY

Bloomsbury Academic
Bloomsbury Publishing Plc
50 Bedford Square, London, WC1B 3DP, UK
1385 Broadway, New York, NY 10018, USA
29 Earlsfort Terrace, Dublin 2, Ireland

BLOOMSBURY, BLOOMSBURY ACADEMIC and the Diana logo
are trademarks of Bloomsbury Publishing Plc

First published in Great Britain 2022
This paperback edition published 2023

Copyright © Scott Aikin and John Casey, 2022

Scott Aikin and John Casey have asserted their right under the Copyright, Designs and Patents Act, 1988, to be identified as Authors of this work.

For legal purposes the Acknowledgments on p. vi constitute
an extension of this copyright page.

Cover image: rudigobbo/Getty images.

All rights reserved. No part of this publication may be reproduced or transmitted in any form or by any means, electronic or mechanical, including photocopying, recording, or any information storage or retrieval system, without prior permission in writing from the publishers.

Bloomsbury Publishing Plc does not have any control over, or responsibility for, any third-party websites referred to or in this book. All internet addresses given in this book were correct at the time of going to press. The author and publisher regret any inconvenience caused if addresses have changed or sites have ceased to exist, but can accept no responsibility for any such changes.

A catalogue record for this book is available from the British Library.

Library of Congress Cataloging-in-Publication Data
Names: Aikin, Scott F., author. | Casey, John, author.
Title: Straw man arguments: a study in fallacy theory / Scott Aikin and John Casey.
Description: London; New York: Bloomsbury Academic, 2022. |
Includes bibliographical references and index.
fallacies, rhetoric, argumentation theory and informal logic"– Provided by publisher.
Identifiers: LCCN 2021035900 (print) | LCCN 2021035901 (ebook) |
ISBN 9781350065000 (hb) | ISBN 9781350065017 (epdf) |
ISBN 9781350065024 (ebook)
Subjects: LCSH: Fallacies (Logic) | Reasoning.
Classification: LCC BC175.A35 2022 (print) |
LCC BC175 (ebook) | DDC 165–dc23/eng/20211014
LC record available at https://lccn.loc.gov/2021035900
LC ebook record available at https://lccn.loc.gov/2021035901

ISBN: HB: 978-1-3500-6500-0
PB: 978-1-3502-8470-8
ePDF: 978-1-3500-6501-7
eBook: 978-1-3500-6502-4

Typeset by Integra Software Services Pvt. Ltd.

To find out more about our authors and books visit www.bloomsbury.com
and sign up for our newsletters.

Contents

Acknowledgments — vi

1 Fallacy Theory and the Straw Man — 1
2 Fallacy Names and the History of the Straw Man — 29
3 Straw Men, Weak Men, and Hollow Men — 53
4 Straw Men and Iron Men — 85
5 The Puzzle of Effectiveness — 113
6 The Puzzles of Dialecticality and Meta-argumentation — 139
7 Consequences for Fallacy Theory — 181

Notes — 215
References — 222
Index — 231

Acknowledgments

This is a book that has been brewing for a very long time. We have been talking with each other about this fallacy pretty much nonstop for over ten years. We have published numerous essays on it, written countless blog posts at *The Non Sequitur* regarding it, and given many conference presentations on it. And we even have been referred to as *the straw man duo* by those who have heard our pitch enough. In this regard, we count ourselves very lucky. We each have an intellectual partner who is up to the task of thinking hard and continuously about something that to so many seems very simple, a fallacy that takes only a page of a textbook to present and a minute of class time for sophomores to get. And here we are devoting our scholarly lives to peel it open like an onion. And we've found a deep complexity to its layers, with connections to other features of argument and an opening to a domain for theory of meta-argumentative fallacies not yet developed. And in this process, we have also shared many laughs and had numerous adventures. We are also very lucky to have friends and colleagues with whom we can try out our ideas on. For this project in particular, these are Andrew Aberdein, Jason Aleksander, Lucy Alsip Vollbrecht, Teddy Bofman, Pat Bondy, Daniel Cohen, Matthew Congdon, Ian Dove, Michel Dufour, James Freeman, David Godden, Geoff Goddu, Lenn Goodman, David Miguel Gray, Leo Groarke, Idit Dobbs-Weinstein, Dale Hample, Hans Hansen, Tempest Henning, Michael Hodges, Michael Hoffmann, Michael Hoppmann, Catherine Hundleby, Gabrijela Kišiček, Jens Kjeldsen, Jan Albert van Laar, Marcin Lewiński, Christoph Lumer, Dan Milsky, Dima Mohammed, Karen Ng, Steve Oswald, Fabio Paglieri, Kathryn Phillips, Cristián Santibáñez Yáñez, Jennifer Schumann, Katharina Stevens, Robert B. Talisse, Paul C. Taylor, Christopher Tindale, Jean Wagemans, Harry Weger Jr., Harald R. Wohlrapp, and Julian Wuerth. We would like to thank Colleen Coalter for initially working out the ideas for this book for Bloomsbury with us, Becky Holland for her work in getting the book over the finish line, and to the team at Bloomsbury for helping produce the volume.

A final word of thanks goes to our wives, Susan Foxman and Katie Casey, for putting up with our many late-night discussions and conference trips, and for their good cheer and support through the writing process for the book. And for their excellent company through all other times, too.

1

Fallacy Theory and the Straw Man

1.1 The Project

It is a common enough sequence of events, especially for social creatures with opinions. You have a take on how things stand or how to fix a problem, but one of your fellows disagrees. Lucky for you both, the two of you share a love of reason. So you agree to argue it out. But here things go off the rails, because your interlocutor, in responding to your reasons, misinterprets them. And it is not just that they are misinterpreted, but these newly considered reasons are just manifestly bad. After noting how bad these reasons attributed to you are, your interlocutor then acts as though the critical discussion is over and that they have shown your take on things unfounded. Congratulations! You've been straw manned.

For as interpersonally frustrating as this experience is for many who are on the receiving end of it, the sequence is a philosophically interesting phenomenon. The straw man is a theoretically robust fallacy. Here are a few features of the fallacy that make it worth deep investigation. First, there's the simple question of how such a move could be effective. In the above scenario, it seems on its face impossible for it to achieve any argumentative success, at least as bringing you around to agree or to move the conversation forward in terms of what's been shown and what hasn't. You won't think you've been refuted, because the reasons refuted weren't yours. Assuming that the core of fallacy analysis is explaining how these argument types can, despite their being instances of bad reasoning, appear to be good and can be effective on an audience, we have the question: how in the world can straw man arguments have any effectiveness in argument? Call this *the puzzle of effectiveness* for straw man arguments.

Second, the straw man is a fallacy that arises not from our reasoning about the things we reason about but from our *reasoning about each other's reasoning*. That is, many fallacies happen when we do our normal first-order reasoning about things. So we can make a hasty generalization about cats, or we might commit the fallacy of asserting the consequent about the gross national product of Finland. But we can commit the straw man fallacy only when we reason about another's reasoning, or when we think about what they were thinking. So, there are fallacies that occur in argument, and then there are fallacies that occur in meta-argument, when we argue about each other's arguments. Because the straw man fallacy arises as a fallacy rooted in our meta-argumentative skills, it seems we must devote particular care not only in its diagnosis but in its correction. It arises because we have developed skills at detecting and diagnosing bad reasoning, so it will be sensitive to the tools we have bearing on bad reasoning relevant to its instances, too. The irony is that it seems that our reasoning about reasoning should yield better results, not a distinct kind of fallacy. Call this complex of issues *the meta-argumentative puzzle* for theories about the straw man.

Third, and finally, the straw man, as an informal fallacy, is unique in its form. At its core, there is criticism of what is well taken to be bad reasoning, or the rejection of what's a manifestly objectionable view. That's a good thing, right? So, why is the criticism of a bad view or bad argument a fallacy? The problem, of course, is that this salutary core of the straw man stands in the way of exchange and evaluations of the reasons held that motivated the exchange in the first place. What's important, then, assuming that arguments must address the standing concerns and reasons that are constitutive of their contexts, is that with arguments we must begin with a picture of our opposition, an objector group. And our reasons must be effective against the reasons they deploy or objections they lodge. So, unlike fallacies of false cause or hasty generalization, which we can commit *on our own*, the straw man requires another person in the argumentative tango. Most other fallacies are solo projects, but the straw man requires others. And second, that even if the criticism of the reasons portrayed is correct, if those reasons we attend to are not representative of what the other interlocutor holds or expressed, then those reasons are not appropriate. Again, even if they are right about the criticized reasons, we still have a fallacy. Call this the *dialecticality puzzle* with straw man arguments.

We think these three orienting puzzles with straw man arguments make them theoretically interesting, and we think that in making explicit how to think them through, we can reveal more about how fallacies work, how our reasoning about fallacies works, and how argument generally has multiple valences that can shift around and change how we see our own reasoning and that of others. That's not nothing. In fact, we think it's quite a lot.

So the straw man fallacy is itself interesting as an object of inquiry. And we, in thinking about the fallacy, have discovered that the straw man itself has a wide variety of presentations. Three forms are distinguishable in the first instance, but we will survey a few further forms that may be special instances of the primary fallacy, but are of unique interest in how they function. First, the *standard* or *representational* straw man is in one speaker misinterpreting another's viewpoint or argument in a way that makes it easier to criticize. Second, the *selectional* or *weak* man fallacy occurs when one speaker selects a weak version of another speaker's argument or viewpoint and criticizes it as though it is representative of the opposition's best case. Third is the *hollow* man, which is the speaker completely making up a set of views for an opponent, attributing them to the opponent and then criticizing them. It is not a *misinterpretation* of anything in particular said by the other speaker but a *complete fiction* woven from prejudices on the side of the speaker. Our plan is to survey these different forms of straw man arguments and show that they should be considered distinct forms, because, first, their presentations are distinct, and, second, the terms for their correction are distinct, too.

Fallacies are themselves an unruly bestiary, if not because they are instances of breaking rules of good reasoning in the first place, then because our categories for argumentation must be wild and woolly as the phenomena we are trying to classify and correct. Consequently, the fallacy of the straw man has a variety of other dialectical instances that are not easily classified within our three classes of *standard straw*, *weak*, and *hollow* man fallacies. They include what we call the *burning* man and the *self-straw* man. We will give these our attention for the sake of showing that the dialectical terrain around strategic representations of one's interlocutors and oneself is varied and often uneven, and so the bounds for straw-man classification are still in the process of being mapped.

A further consideration regarding the straw man is one familiar to most in fallacy theory—whether there could be argumentatively appropriate instances of straw man arguments. So, for example, there are arguably appropriate instances of *argument from pity*, as invoking the emotion of pity can be, under the right evidential conditions, a relevant moral reason for a conclusion. For instance, appreciating the suffering of veal calves, with their space and contact with other animals restricted, their diet being entirely of milk, and their consequent propensity for disease, can be a relevant reason to oppose the production and consumption of veal. Being emotionally moved by suffering is an indicator of a morally relevant consideration being presented. So, not all arguments from pity are fallacious. The question, then, is now posed about the straw man—can there be argumentatively useful instances of straw man arguments? Our answer is *yes*, and in fact, we hold that there are forms of the straw man that have salutary instances, particularly in clarifying a dialogue and its progress.

But the possibility of non-fallacious distortions of an opponent's view raises another thorny question: whether, instead of constructing *straw* opponents, there are relevantly similar considerations bearing on constructing *iron* opponents. That is, the metaphor of the *straw* man argument is that one has strategically made one's dialectical opponent's view easier to criticize by interpreting it as a less plausible thesis or argument than was given. One's opponent is made of *straw*, something flimsy and certainly not permanent. But, on the metaphor of *iron* man, one *improves* the opponent's position, interpreting and presenting it as a more plausible version of the thesis or argument given. For sure, there are salutary epistemic benefits that can arise from iron manning. To start, in taking up with the best versions of interlocutors, we display a good will and project cooperative argumentative spirit, and it makes it clearly more likely that higher-quality output is the result of the exchange—it will not only be a more stable agreement but it will more accurately approximate the truth of the matter. It is certainly clear that iron manning is good in many circumstances. But is it good in *all* cases? We believe it is not, and so propose the form of the *iron man fallacy*. It can take three forms. First, there can be the problem of moral hazard with iron manning—it may make our interlocutors dependent on our interpretive charity and not do the work of improving their own views. Consider this as a particular problem with grading student papers. Second, there is the third-person straw man problem for iron manning. That is, we are often not the only interpreter and critic of another's argumentation, so an iron man of one speaker's contribution may straw man another's criticism of the original (less plausible) statement. Iron manning yields inaccurate argumentative scorekeeping. Third, and finally, iron manning's desideratum of improving another's contribution in interpretation can have unacceptable silencing of voices from the margin, as the *plausibility* of a view or reason is itself a matter of debate and controversy. So if there are broader disagreements, iron manning the opposition's views may misrepresent the depth of the disagreement that separates the sides. Sometimes, charitable interpretation is a way of talking over the other. So we must view the iron man as a generally virtuous tactic of argument, but it, like all tools of virtue, must itself be performed at the right time, to the right degree, in the right manner, and with the right audience and interlocutors.

1.2 Fallacy Theory: A Defense

Our work on the straw man here is a contribution to the field of fallacy theory. Fallacy theory is the convergence of three broad programs in the study of argument. First is the first-order research program of defining fallacy, finding and taxonomizing new types. Second is the pedagogical program of teaching some taxonomy of fallacies as part of critical thinking classes. Third is the meta-theoretical program of articulating what the relationship is between understanding fallacies and the broader program of understanding arguments and reasoning. We'll note that this book is primarily a contribution along the lines of the first and third domains of fallacy theory (working out new types of fallacy and then turning to talking about what it teaches us about argument more generally), but we'll comment pretty liberally on the second domain, that of teaching. In fact, a number of our best examples come from the classroom and our roles as educators.

Fallacy theory has come under considerable criticism of late, along all its fronts. There are three rough classes of objection. First is the *generality problem*: fallacy theory's generality loses its connection to actual argument instances, but its particularity loses normative bite. Second is the *scope problem*: it seems the number of fallacious forms is unlimited, so there is no well-defined domain of study. Third is the *negativity problem*: foregrounding failure and the vocabulary of criticism promotes argumentative adversariality, and as a consequence contributes to bad argumentative practice. Again, given that we are committed to doing fallacy theory (because, you know, we're working this book out about the *straw man fallacy*), it's worth answering these problems for the broader program, and we will explain in particular why it is useful for the theory of straw man fallacies.

We here offer a brief reply to these challenges, and we'll turn to deeper answers in a few pages. To the *generality problem*, our reply is that fallacy theory provides a vocabulary for critical evaluation in discussion. All normative vocabulary will have some version of the generality problem, but this problem actually calls for more refined vocabulary, not the rejection of it. And so this is why we have a wider program of kinds of straw man fallacies. To the *scope problem*: the fact that there are more varieties of fallacious argument is good news for the discipline, not bad. In fact, one of the consequences of doing fallacy theory is that, in doing it, we make argumentation more reflexive. That's good news in a way, but that reflexivity makes new fallacies possible, which is bad news. But we can manage this problem. We'll give a survey of those thoughts here in Chapter 1, but we will return to the big idea behind the reflexivity of fallacy theory and argument in Chapter 7. To the *negativity problem*, the reply is that argumentative exchange is best conceived as *dialectically minimally adversarial*, and so fallacy theory must then provide tools for articulation of criticism and also the tools for management and de-escalation of critical discussion. The straw man simply can't be understood except under conditions of adversarial argumentation, so it's worth getting clear what precisely makes things opposed in critical dialogue in the first place.

Our plan here is to briefly survey what we see as the three domains of fallacy theory, then turn to what we take as the three main programs of criticism. Finally, we offer our modest defense of fallacy theory. By this we mean that we concede much of the critical bite of the cases against fallacy theory but hold that these are welcome occasions for reform and reconception.

1.2.1 Fallacy Theory and Its Components

Fallacy theory is a subdomain of argumentation theory or informal logic, which is a research program devoted to the study of argumentative normativity. A commonplace is to contrast the focus of this broader domain with that of formal or deductive logic; the latter concerned with conditions for argumentative validity and the former concerned with the weaker forms of support for arguments as products and other procedural issues with argument as process. As far as it goes, this is a useful estimation of the broader domain, and fallacy theory is the more restricted study of ways support fails or procedural rules of argument are broken. Exactly how to even thematize these failures is precisely one of the core issues of fallacy theory. We all know that fallacy theory is a subdomain of informal logic focused on argumentative failure, but what constitutes that failure is a matter of debate within fallacy theory. And so, there are divisions about how to even define what a fallacy is. There is the "standard treatment," as identified and criticized by Hamblin (1970), that fallacies are arguments that seem valid but are not.[1] There is the broadened version, as developed by Johnson (1987: 246), that a fallacy is an argument that violates one of the standards for good argument and occurs with sufficient frequency to merit being classified. And the pragma-dialectical perspective, as seen with van Eemeren and Grootendorst (1987: 297), that fallacies are discussion moves that threaten the resolution of a dispute—and in particular, they are violations of rules of critical discussion. Alternately, a fallacy may be, as Walton (1995: 15) terms it, the misuse of an argument scheme. There are, of course, more varieties of definition, and they generally depend on the theory of argumentative normativity on offer, as all theory of fallacy is a theory of how one *fails* to do what one *ought* in argument. Disagreement about argumentative norms yields disagreement about what it is to break those norms or fail their demand. That's just how theory and its application work, isn't it?

The second focus for fallacy theory is about how informal logic is taught in the classroom. Again, a contrast with formal logic is useful. With natural deduction, the focus is on rules of good inference and their systematicity, particularly in construction of proofs. Little systematic effort is put into the articulation of ways to fail at the objectives of proof except when giving tips on how not to mess up those derivations in the homework (e.g., don't forget to discharge all your assumptions of conditional proof in reverse order of their introduction!). In contrast, the overwhelming amount of time and energy put into classroom work in informal logic is on fallacies—how arguments fail. And so training for students is often in the form of fallacy-spotting, not argument construction. Work in fallacy theory informs pedagogy in the sense that well-taxonomized and explained accounts of fallacy allow students a rich interpretive framework for discussion. The objective of looking for "fallacies in the wild" from the pages of newspapers and out of mouths of talking heads regularly yields substantive class discussion.

The third, metatheoretical, component of fallacy theory is the task of articulating how findings in fallacy theory inform our broader research into argumentative and discursive axiology. What does a certain fallacy reveal about argumentative norms? What does the prevalence of a class of vicious dialectical tropes tell us about our society? How does argument, even though we are regularly bad at it, fit with democracy? A natural thought is that certain argumentative failures are pregnant with meaning about what argument should be, how it should work. And so, out of a few object lessons in how not to argue, we have information about how to argue. And so, a kind of reflective equilibrium arises between our theories of argument and our systematic treatment of fallacy. Well, at least that's the hope.

We will next present the three main families of objections to fallacy theory, and one thing we think arises from their presentation together is a familiar picture, at least to philosophers. It is a picture of a domain of study that has as one of its central and most fractious issues the question of what it is about and whether it is worth doing at all. Ever since Thales had his pratfall in the well, philosophers regularly have had to answer these questions for themselves: what are philosophy's objects? What are its standards? Is an education in it a hindrance to being a useful human being? And so it goes for fallacy theory.

1.2.2 The Generality Problem

The generality problem for fallacy theory is an instantiation of the wider problem of how norms govern particular actions. Norms are general, if not universal prescriptions. Yet particulars, insofar as they are particular, never are mere instantiations of universals, but are always roughly classed as such.[2] Many the nominalist has said universals are mere words. So a version of this challenge arises for fallacy theory. Maurice Finocchiaro captures the thought: "[T]here probably are no common errors in reasoning. That is, logically incorrect arguments maybe are common, but common types of logically incorrect arguments probably are not" (2005: 113). In a similar vein, Boudry, Paglieri, and Pigliucci pose what they call the *Fallacy Fork*:

> [O]n the one hand, if fallacies are construed as demonstrably invalid forms of reasoning, then they have very limited applicability in real life (few actual instances). On the other hand, if our definitions of fallacies are sophisticated enough to capture real-life complexities, they can no longer be held up as an effective tool for discriminating good and bad forms of reasoning.
>
> (2015: 431–2)

Massey, similarly, charges that the "myriad and intricate schemes for classifying fallacies suggests there is little behind the science of fallacy… [T]here is no theory of fallacy whatsoever" (1981: 491). At best, fallacies are "subjective" and at worst they are empty, as there are simply no instances of them beyond what occur in textbooks. We, ourselves, have joked that our work in fallacy theory is more driven by our pet peeves about how others argue (and our concomitant desire to make up funny names) than a systematic theoretical orientation. Well, that was the joke, but now it's not so funny.

For sure, the history of fallacy theory is testament to the fact that it is usually an *ad hoc* repository of pet peeves of intellectuals about the linguistic behavior of others. We aren't the only ones! Aristotle's *Topics* and *Sophistical Refutations* certainly read as such. And the current work in fallacy theory in developing new vocabularies of dialectical criticism is for the most part reactionary scholarly work of seeing patterns of argumentative vice in the buzzing blooming confusion of public reason-exchange.

The strong reply is given by Johnson—that we should eliminate the subjectivity of fallacy theory in the same way that logic should resist psychologism: fallacy theory "should be purged of its subjective and psychologistic nuances" (1987: 245). And so the "seems" talk of fallacy theory (that of taking a fallacy to be an argument that only *seems* valid but isn't) is to be eliminated.

But this is much too strong a solution, if only because we want fallacy theory to do some double-duty in theorizing fallacies. That is, we want to not only (a) explain why the argument given is not good, but also (b) explain why people continue to give it and people accept it, too. This is why *seems* itself appears ineliminable in fallacy theory: we want to be able to explain why we fall for certain bad arguments, and "seems" talk is the best vocabulary for illusions of argumentative quality. This double-duty program, further, is the reason why fallacy theory must invoke types in order to evaluate the tokens of the fallacy forms. The type instantiates an argument scheme that, given informal factors, will be appropriate or inappropriate. And the type, given the way we (or many) happen to reason in particular circumstances, appeals to us in particular tokens. The hope is that if we can explain how a bad argument could be compelling to someone, we hope to make it so that, maybe, it won't have that temptation over us. That's how we break the spell of fallacies when we see them before us, and that's what we are hoping to do theoretically, too—figure out why we fall for such nonsense and, in figuring it out, keep it from continuing to happen.

However, the talk of seeming points to another long-standing complaint about fallacy instruction and theorizing: the examples that populate the exercise sets of innumerable textbooks and frequently form the basis of discussions in the scholarly literature are very often too ludicrous to be credible or to be of any value. They don't *seem* to be good at all! They, rather, seem manifestly bad. Finocchiaro (1981/2005) writes:

> [T]he examples of the various types of fallacies, [are] usually rather meager. It consists mostly, if not exclusively, of more or less artificially constructed examples for the purpose of illustrating the various descriptions of fallacies. Examples of fallacies actually occurring in the history of thought or in contemporary investigations and controversies are rare.
>
> (14)

A similar worry can be found in Boudry, Paglieri, and Pigliucci, (2015):

> The problem is that one hardly ever encounters "fallacies" in this guise. They only appear as highly artificial textbook cases, designed to be knocked down easily.
>
> (435)

It's certainly true, as Walton (1989a: 171) also notes, that textbooks and even essays on fallacy theory are rife with context-free tidbits of arguments meant to stand as tokens of the various fallacy types. We've written scores of these ourselves for tests and quizzes, and the whole objective of showing them to be fallacious is that they need to be manifestly bad. What is more troubling, so the objection runs, is that people *theorize* from the fictional examples. Many are the discussions in the scholarly literature of completely made-up cases of fallacy. We're guilty of this too. And, for reasons we are about to explain, we will continue to be guilty of it.

But there is more to this point than an objection to fake examples in place of real ones. For it is alleged that the near-universal reliance on fake examples is evidence of the unavailability of real examples. Real examples, after all, would be preferable, and fake examples are, well, fake. The reason, so the argument of the Fallacy Fork goes, that real examples are practically unavailable is because they never quite fit the theory. An example of this is provided by Boudry (2017). He laments that continued attempts to have students track down fallacies in the wild, as it were, showed just how pointless a task this is. The problem is that textbook fallacy treatments, even those that strain to rely on "real" examples, deprive the supposed bad arguments of actual context and so they're never quite as bad as you'd think in real life. The conclusion, then, is that given the absence of real examples, and the proliferation of fake examples, it might appear that fallacy theory in theory is rather like doing astronomy from astronomic charts in *Star Wars* movies.

In reply, we'd like first of all, since we're being modest, to concede that much of this is right on the mark. We've tried ourselves the exercise Boudry mentions with similar results.[3] And the method of tracking down fallacies in the wild or playing "pin the tail on the fallacy" is pedagogically and epistemically impoverished (for reasons we will discuss in Section 1.2.4). Having said this, we'd now like to offer a modest defense on behalf of the fake. The defense is in two parts. In the first part we discuss some of the reasons for relying on fake examples. In the second part, we discuss the deeper questions this question provokes. We're going to use some fake arguments in this book. There are pedagogical, political, practical, and philosophical reasons for this.

Fake arguments are often *pedagogically* necessary: When we first started teaching critical thinking, we tried to keep it relevant and applicable by using examples from the newspaper opinion page (where you get most of the arguments). We even started a blog to catalog them. Yes, we're dating ourselves because now there is social media, but this was then. There, you faced a choice: the "letters to the editor," written by ordinary citizens, offered a bountiful harvest of fallacious arguments. The opinion editorials were a step up, though, we assure you, they too were rife with fallacies of every stripe. As up-to-the-minute and empirically grounded as this approach was, it suffered from a few problems.

The first problem is that much context is unavailable when we turned to the classroom. This context doesn't necessarily make the arguments any better, mind you. It rather shows how bad the arguments were. The problem was that most of the arguments, to make much sense, required lots of context. But the work it takes to provide the context takes time away from the main lesson: the logical mistake on display. A second problem is a variation of what we call the *Anna Karenina problem* for fallacy theory: happy arguments are all alike, but unhappy arguments are all unhappy in their own peculiar way. Or the mistakes were imperfect exemplars of the type of error you're trying to show. So something would hover between a straw man and an *ad hominem*. Now, of course, the concepts of fallacies are heuristics (more on that shortly), so this is an important lesson, but it again muddies the waters when the point of the lesson is to introduce something. It makes little sense to introduce a topic by talking about *exceptions*. A third problem is that students were inclined to be either too charitable or too critical. So much of that depended on whether they antecedently agreed with or disagreed with the target for evaluation. They either instantiated what we've elsewhere called *the problem of agreement* or they exercised what one might call *the hermeneutics of antipathy* with the letters, and either way, we found that teaching critical thinking usually turned into an exercise of teaching the skills of rationalization. Fourth, using letters to the editor seemed at times a little mean. While the cranky uncles (full disclosure: we are they) writing letters to the editor over this or that minor outrage are public documents and worthy targets for criticism, it seemed a little mean to pick on people in their absence. Not quite the democratic civic spirit one would hope for with people reading the statements from one's civic equals. Finally, the op-eds tended to be difficult reading—they were not quite as accessible as we would have preferred, since many weren't well written or there's too much inside baseball. In a way, they were bad examples all-around for our sophomores. One reason was the context reason we just discussed. The higher level of the context just made it all the more difficult.

Now for an analogy we'll use later in the book. Not only is this book about straw man argument, it's a book that employs straw man arguments because, as we'll be arguing, you can't argue without them. They're necessary. If you're like us, you've probably tried to learn a musical instrument. The first pieces of Bach or Slayer you learn are simplified versions to test your chops on. They're trimmed of the flourishes and ornaments, and perhaps even their time signature is altered, and some parts are left entirely out, so that you can practice that little piece of music theory you're prepared for. If you show up to the Bach competition with your learner's version of the *Well-Tempered Clavier*, you are going to be in for a surprise. Though with Slayer, just play it fast and loud, and you'll be fine.

What's at stake in fallacy theory is that the primary explananda are not token bad ideas or false claims, but token bad *inferences*. This is a *challenge* for fallacy pedagogy and theory as tasks that are by their nature normative, rather than an *objection* to them being done. The challenge is to capture bad inference schemes without necessarily adding all of the context. If anything, the context (see above) can stand in the way of revealing the argument scheme. Rocky Rabbit's straw man argument is therefore less distracting than Donald Trump's (though we will have a few of those, too).

A philosophy pedagogy version of this argument is an even more poignant appropriate reply to the challenges leveled by Finocchiaro and Boudry et al. Brian Ribiero (2008) notices that teaching philosophy practically requires straw men. Sure, the best version of the ontological argument would be nice, but this is Introduction to Philosophy, and the students are not ready for that. Heck, most of the grad students at the departments where we got our degrees weren't ready for the best version of the argument. And if you hold your sophomores to the standard of accuracy and rigor of professional philosophers, no one will learn anything and, what is worse, no one will major in philosophy. Oh, and don't forget your student evaluations—your deans won't let you, for sure.

A second major argument on behalf of the fake is that they are often *politically necessary* or at least politically *expedient*: argumentation instruction and theorizing involves types. But types are illustrated in tokens. The problem is that people bristle at examples of mistaken arguments from their own point of view. They are inspired to iron man them; they fill in the context that is not dialectically obvious. So the real examples backfire. Besides, people make the meta-mistake about the domain of the instruction with such examples: if they are not balanced by apparent political affiliation, they conclude that the author or instructor is guilty of bias and is subtly trying to show the weight of evidence favors their view. Turning a critical thinking class into one pushing a preferred ideology has drawbacks, for sure. But here's another side to this: attempting to balance out examples for political/pedagogical reasons risks committing the meta-fallacy of bothsiderism (see Chapter 7), if only by implication. Students may draw the erroneous conclusion that mistakes are evenly distributed among all parties to a dispute, or, what is worse, that it is always virtuous to match the misdeeds of one side with those of another. Call this *whataboutism* (again, see Chapter 7).

Though perhaps these are not independently sufficient, there are further strong practical reasons in favor of employing fake examples: audience and economics. By accident of geography, we find ourselves teaching at universities in The United States of America. The textbooks we use are written, largely, for this audience. A quick look at text written for say a Canadian audience, whence many of our esteemed colleagues hail, is enough to make the point: tales of the Canadian Parliament are lost on readers from the United States and so they lack the same punch.[4] Textbook companies have economic reasons to favor content that is geographically expansive and durable through time. Dated examples also suffer from the same problems as geographically specific ones. Zingers from the Reagan (or now Clinton, Bush, or even Obama) era in the United States just do not have the same edge. Similar reasons can be advanced for fallacy theorizing. Argumentation is an international discipline with a moving target, and real examples just go stale.

So much for our reasons for employing—we'd stress, with abundant caveats—our made-up fallacy examples. If these have not been convincing, we have another argument. The core of this thought is this: try as you might to avoid the appearance of fakery, you'll never be successful. For underlying the Fallacy Fork dilemma is a central paradox of fallacy theory generally. Irish logician Richard Whately (d. 1863) seems to have entertained this exact objection. In a rather lengthy section of his *Elements of Logic* (1826), he notes how difficult it was to find examples of fallacies that would be believable *as fallacies* to a consumer of a text on fallacies. In the first place, fallacies taken out of the context in which they actually appear and restated wouldn't be very convincing. The context in which they appear is rather more obscure, which is, of course, the point.[5] It's not going to be a very good trick if you can easily see it. For this reason, fallacies will be "obscured and disguised by obliquity and complexity of expression" so they may the more easily "slip accidentally from the careless reasoner, or to be brought forward deliberately by the Sophist" (p. 186). Context, of course, is crucial in obscuring the presence of the fallacy, in making it more palatable. But there is a deeper point. Trickery, of which fallacies are meant to be exemplars, poses a kind of epistemological paradox. The many examples gathered in fallacy compendia are so many convicted criminals: "they are in fact already detected, by being stated in plain and regular form, and are, as it were, only brought up to *receive sentence*" (1826: 187). As Whately observes, fallacy examples appear to be silly for the simple reason that texts are teaching about fallacies, so the reader is primed to spot them. Seeing them in all of their obviousness as bad arguments is a testament to a lesson learned or perhaps a lesson not needing to be learned. This would be a bit like saying a magic trick was not very deceptive after a discussion of the mechanism of deception; the jig is not only up, but it was up before the trick even got started. The very nature of fallacy pedagogy and theory is corrective in the same way. The examples are silly because the audience in the textbook is not the actual target audience for those fallacious arguments. Authentic examples, of the kind that might convince a skeptical theorist, thus elude detection because of the very (theoretical) circumstances of the discussion. This problem generalizes. As the objection indirectly shows, fallacy theory pursues an objective that is always one step ahead. What is essential to fallacies, so our modest reply goes, is the way they mimic not only reasons but good reasons. Now, of course, they are not actually good reasons. In retrospect and on reflection, they look hilariously bad. The problem is, as Whately notes, that one never has access to the feeling of being hoodwinked by bad reasons because bad reasons are still reasons of a sort. And feeling the tug of a reason you also see as bad is the problem—seeing it as bad undoes the tug, and feeling the tug means it seems good to you. That's the paradox. Fallacies are therefore painfully difficult, if not just impossible, to observe in their natural environment: the mind of someone who believes them.

It is also worth noting that there is an interesting paradox to this objection. The objection runs that fallacy examples are silly arguments no one actually makes. Put in our preferred vocabulary: fallacy examples are *hollow men*, and, as a result, they're *easy to knock down* and so *prove nothing*. Weighing this objection reveals that there is value in identifying token errors (fallacy theorizing) by their type (making stuff up, i.e., hollow men). Now, of course it is true that the fallacies used in theorizing are hollow men. This is just one of the things that is interesting about the straw man fallacy as a family of related errors (and one of things we'll talk about later). And one of the great ironies is that *we ourselves have been charged with weak- and hollow manning* with our examples in our previous papers. Of course, we have to make up whole cloth a terrible argument (as is a hollow man), and, of course, we go out to pick out the worst instances of arguments (as is a weak man), because we are looking to theorize bad reasoning. But because we also think that there are salutary instances of these argument schemes, too (as we show in Chapters 3 and 4), this is not the problem it seems. But it is a complication.

The modest reply, then, is that fallacy theory comprises of a program of identifying groups of argument types that have rough family resemblances among their tokens.[6] Notice that this rough notion of the normative category of fallacy is analogous to the moral categories of criticism. *Negligence* is a property of a class of actions one might find objectionable, and though negligence takes many forms, there is still the rough notion of *could have and should have been more attentive* that comprises the class. So too with *theft*, *mendacity*, and *selfishness*.

The generality problem, as posed by Boudry, Paglieri, and Pigliucci's *Fallacy Fork*, certainly captures the problem of critical vocabulary being either too specific to be statements of rules or too general to have any obvious connection to real-life arguments. But consider that once one learns the language of fallacy challenge, the charge of fallacy is part of what Johnson calls "initial probing" in critical discussion. Fallacy theory is the development of a meta-language for argumentative criticism, and learning the language of fallacy theory is not just that of making fallacy charges but of putting nuance on an argument, requiring clarification, developing in a critical discussion.[7] The reality, of course, is that vanishingly few actual arguments come in textbook fallacy form. This both fallacy theorists and critics of fallacy theory avow. But that does not make the language of fallacy-assessment irrelevant. When it can be plausibly charged that an argument instance is a token of some fallacy type, then it is incumbent on those who either give or accept it to defend against the charge—some relevant piece of evidence is brought to light or the clarification of a connection may arise, or they may have little or nothing to say. The point is that fallacy vocabulary and the theory behind it is in the service of the reason-and-challenge structure of argument exchange. As a consequence, the generality problem is both bad news and good news.

The generality problem is bad news in the sense that, because our evaluative categories must instantiate general normative outlooks and specific instance emphasis, they will have penumbral edges and overlapping cases—this is why it is often so difficult to classify some bad argument instances.

The good news is that in the general forms fallacies take, fallacy theory can provide explanations for why some argument correction can be so difficult and can provide roadmaps for critical discussion in light of how the arguments are challenged and so on. The phenomenon described is itself complex and variegated; so, too, must its description. Thus, a *modest* reply to the generality problem.

1.2.3 The Scope Problem

The scope problem is simply that fallacy theory has no clear demarcational criteria—what distinguishes fallacy theory from culture criticism, political philosophy, or rhetorical analysis? In light of how the modest solution to the generality problem was that fallacy theory has, as a matter of course, rough terms and open-ended evaluative criteria, the scope problem for fallacy theory becomes all the more trenchant. It seems, on its face, that the modest fallacy-theoretic reply to generality amplifies the scope problem that there is no clear limit to what fallacy theory should evaluate; it is, by its own description, an incomplete task.

For example, Catherine Hundleby observes that the domain of most fallacy theory is from the perspective of those who are roughly social equals trading reasons with other social equals. This, for sure, is a relevant domain, but it is not exhaustive of the scope of bad and recurrent argument types. Hundleby observes that too much is left out—there are "androcentric fallacies" (2009: 2), and there is a growing literature on how too many from underrepresented groups are not given their due in critical dialogue, beginning with epistemic (Fricker 2007; Medina 2013) and extending to argumentative (Bondy 2010; Henning 2018; Hundleby 2013; Rooney 2012) injustices. Moreover, even within the standard model of fallacy theory, it seems that given the modest solution above, there could be, for every fallacy, particular sub-instances of the fallacy (as seen in our multiplying versions of "straw man fallacies").

From a purely theoretical perspective, the scope problem is really nothing but good news—the variety of argumentative error and systematic explication is wider than anticipated and seems (assuming the examples above are right) to bear significant connection to pressing social issues of the day. Fallacy theory, like critical thinking more broadly, finds its way into anywhere we humans reason (or at least purport to do so well). Consequently, there is much for the fallacy theorist to do: new vocabularies to devise, new strategies of argumentative repair to develop, and programs of connecting these to standing taxonomies of fallacy and argumentative theory. The scope problem, from this perspective, is just a manifestation of the *fecundity* of the research program.

The bad news is the return of the generality problem. How *applicable* is this progressively more complicated machinery of argument evaluation? Recall that the modest solution to the generality problem was to argue that the intersecting and penumbral categories were part of all normative-evaluative discourse, and so managing these categories requires regular application of principles like those of conceptual tolerance (that some concepts are *gradable* and *intersecting* instead of *absolute* and *exclusive*). The scope problem can be restated now as a special instance of the generality problem: *fallacy theory, in adopting penumbral, intersecting, and expanding categories for fallacy, is on the way to an impossibly unwieldy classificatory and critical system.* In pursuing the specific-critical line in instances of bad reasoning, the general program of identifying fallacy tokens of types has been foregone. The modest reply, then, is that of identifying the objective of developing a meta-language of challenge for reasoning.

The phenomenon of challengeable reasoning is complex, and so the vocabulary will be, too. How fine-grained the vocabulary is depends on the scrutiny to which we are subjecting the reasoning (or how dialectically deep the challenges and replies have gone), and how many new fallacy forms we wish to introduce depends on what, precisely, the critical project is out to accomplish. The language of logic is an instrument of rational self-consciousness, and depending on what needs to be made explicit, the vocabulary's detail can be brought to bear or forborne. And so, this return to the modest solution is *modest* in that it concedes a good deal of the scope problem. The objective of fallacy theory must be defined and refined against the issue or problem's backdrop of what the points of argument-analysis in the contexts are.

An analogy here may be useful. For anyone who has taught formal logic or uses its tools with regularity, there is the question of just how large the toolkit of rules of implication and equivalence should be. Some like their toolkits lean and mean, so only a few necessary rules and then a tolerance for longer ways around with derivations. Others are just fine with great lists of rules with overlapping application, as one's derivations are then more elegant and direct. And then we see the attempt to find some median that provides sufficient breadth of rules to move proofs along with some intuitive pace, but does not overburden with the sheer number of valid forms. And so, one asks oneself, *do we really need constructive and destructive dilemma*?[8] In some ways, the task of deciding here is that of trying to identify just what one is trying to do with both the logic as formulated and in its application. We are looking for both a satisfying theoretical approach to not only detect but explain good reasoning, but we also want it to be something that the average sophomore can master in a few months and apply to some reasoning here or there. Nobody ever said that because the scope of valid forms (and so potential rules for logic) is not bounded, it's not clear what the bounds of formal logic are. That would be a pretty silly thing to say, really. Rather, the point is that we scale for our objectives—what needs to be detected, what needs explanation, what we can expect from the sophomores. The same goes for the bounds of fallacy theory. Really, it's all *non sequitur* or *ignoratio elenchi*, right? (More on that point in Chapter 2.) If that's the case, then why have any more forms for bad reasoning than those two? The answer is that, yes, we could probably do it all with one generic notion, but in order to explain errors in reasoning with the level of particularity we are looking for, the varieties must be represented and a set of categories must be arranged. And once we do that, the distinctions start getting made pretty quickly. Why? Well, as it turns out, there are just a lot of ways for reasoning to go wrong.

All this said, there is an improved version of the scope problem. (Note, by the way, how we are *iron manning* an objection to our project.) The scope problem is that the variety of bad reasoning and inappropriately performed argument is wild and woolly, so fallacy theory will yield a too-unwieldy theoretical overhead for it to be both useful and accurate. Our answer was that, given particular needs, the theoretical tools of the theory make the practice reflexive, criticizable, and correctable. That's good, sunny, hopeful theory. But anyone who's taught a critical thinking course knows that there is a darker, more pessimistic outlook possible here. Students, once they learn the fallacies, just love to spot them and charge others with committing them, and so expect that dropping some Latin name for an opponent's argumentative move will be a kind of magic that fixes the critical discussion (often in their favor). Call it the *Harry Potter problem*, thinking that learning a little Latin phrase will paralyze an opponent with its expression. So entirely new kinds of wild and woolly behavior are encouraged with this vocabulary. Without fallacy theory teaching the fallacies, there would be no burden of specious allegations of fallacy. Further, new fallacies are made possible—take for example, the *fallacy fallacy*, inferring that an opponent's position is false because it is supported by fallacious arguments. Again, this move is made possible by the development of our normative vocabularies as corrective tools, but it turns out that the tools are capable of being misused. And so we, in making our practices more self-reflective, create opportunities for not only new corrections to first-order pathologies in those practices, but create new opportunities for second-order pathologies that arise out of our corrective reflection.

Aikin has elsewhere (2020) termed this general phenomenon *the Owl of Minerva problem*—the Owl of Minerva flies only at dusk; wisdom comes only in hindsight. The general form of the problem is that we, as reflective creatures, make new concepts to improve ourselves, but since we are aware of those concepts, we change our behavior in light of them and our knowledge of them bearing on us. Ian Hacking calls this a "looping effect" that concepts can have—they, in being about things that are aware the concepts are about, end up changing the things they are about. The concepts and the things they are about interact. We are like that, and many of our concepts work that way—the concepts don't just describe us but seem to, as we act, give our further actions new direction (1999). Think of how in your own life terms that bear on you suddenly don't just *describe* you but *direct* you, once you see them as being about you. Intellectual, nerd, jock, liberal, and so on. That is, they aren't just true about you, but now you reason along the lines of our identity, "I'm a liberal, so I'll … ."

The thing is that self-awareness can yield good changes and clarity with our practices. That's, for example, what the concept of a "foul" does for sports—it is supposed to make it clear what is acceptable and isn't in the practice, and we then play in light of that knowledge. And so the same goes for fallacy theory: the concept of fallacy is supposed to work roughly like the concept of a foul, but in the game of giving and evaluating reasons. But the concept of fouls in sports yields inappropriate uses of the concepts—for example, the gamesmanship of *diving*. When one *dives*, one simulates being fouled for the sake of having the penalty for the foul assessed against the other team. So one gets free throws in basketball, a penalty kick in soccer, a walk to a base in baseball, and so on. That's really a kind of meta-foul. It's a pathological looping effect in sports, and it's because we have that meta-awareness of the practice and the relevant concepts that comprise that awareness. Most of what we do and teach in fallacy theory is meta-argumentation. But here's the catch—just as there are fallacies in first-order argument, there are fallacies in meta-argument. In fact, straw man arguments are a prime example of meta-argumentative fallacy—they are pieces of bad reasoning that arise from our reasoning about (another's) reasoning. And so, we get meta-argumentative fallacies like the straw man, the fallacy fallacy, the fallacy of bothsiderism, the fallacy of the truth being in the middle, fallacies arising from enforcing norms of civility, the free speech fallacy, and in philosophy the fallacy of what Aikin and Talisse call "metaphilosophical creep" (2018: 135). We will have a lot more to say about this phenomenon, the new kinds of fallacies, and how to manage the problem in Chapters 6 and 7.

One lesson is that the Owl of Minerva problem with fallacy theory is a strengthened version of the scope problem—in this case it's fallacy theory and its self-conscious application itself that exacerbates the problem of scope. Now, that's a *hard* problem! To be honest, we don't think there's a *solution* to the Owl of Minerva problem generally or to the specific version of it for fallacy theory. The reason why is that whatever solution it would be would, for it to be effective, have to be accessible to us as self-reflective creatures. But because of the looping effect, the solution itself is open to pernicious error in application. A case in point is that when we identified the fallacy fallacy, we'd not closed the issue on meta-argumentative fallacies, because there is still the case that it's possible to commit the fallacy fallacy fallacy about fallacy fallacies.[9]

The key take-away from this discussion is that the Owl of Minerva problem is not just a significant issue for fallacy theory but for most any normative theory with the hopes of getting the theory right accords with managing the practice—just look at how ethics provides too many with the tools for rationalizing their choices, instead of properly directing them. And, again, we don't think it is fixable. So what, then? The first big point is that the cat's already out of the bag with fallacy theory, so the Platonic plan of not letting the *hoi polloi* know about the theoretical view on them is not viable. The word's out, and the looping effect is already at work. So *stopping* the dispersal of fallacy theory is unlikely to fix anything, and moreover, it's not really a good sell for our critical thinking classes. Deans don't generally go for the news that teaching the students these norms will create the possibility for totally new metatheoretical pathologies in the practice. But, again, here we are. The better attitude to take here is that of finding ways to *manage* the problems that come with having these programs. So just as it's not a reason to stop doing ethics if doing ethics allows us to rationalize more effectively, the fallacy fallacy is not a reason to stop with fallacy theory. What's necessary is that we develop new tools for detecting and correcting the meta-argumentative pathologies our last generation of corrections gave birth to. So the bad news is that we have practices that need to be improved. The good news is that we can do that with making the norms of those practices explicit to their practitioners, and the second piece of good news is that we do that pretty well and have good heuristic meta-vocabularies to self-consciously improve those practices. The follow-up, and new, bad news is that those meta-vocabularies also need to be monitored for their inappropriate application and misuse. The hopeful news (but not yet good, we admit) is that we are developing vocabularies to fix those meta-level errors, too (more on this in Chapter 7). And so on it will go, but that's the fate of fallible self-reflective creatures like us, isn't it?

1.2.4 The Negativity Problem

Fallacy theory is a systematic view of argumentative *error*. The vocabulary of fallacies, as a consequence, is univocally *critical*. There are two consequences of this negative-emphatic view. The first is that fallacy theory has a *problem with misplaced emphasis*—we should be looking for ways not only to criticize arguments but to construct good ones and improve bad ones. The second is that fallacy theory, in its negativity, is complicit with (and in many cases actively promotes) the excessive adversariality of argumentative exchange.

Catherine Hundleby (2009, 2010, 2013) and Phyllis Rooney (2010, 2012) have argued along both of these lines. Because of fallacy theory's negative valence, negative consequences ensue. Hundleby observes:

> The oppositional nature of fallacy-allegation... lends itself to formulations according to the politically regressive and epistemologically archaic Adversary Paradigm
>
> (2010: 280)

Hundleby further observes that the way fallacies are regularly presented in textbooks offer "no suggestion of argument repair" (2010: 289) and yield "pin the tail on the argument" exercises for students. Phyllis Rooney, similarly, argues that the adversarial paradigm is epistemically and argumentatively stunted:

> [T]he Adversary Paradigm either leads to bad reasoning... or... it sustains a more limited range of reasoning and argument forms.
>
> (2010: 205)

In short, the negativity of fallacy identification is part of and contributes to the Adversarial Paradigm, which obscures the goals of truth-seeking.[10]

The modest defense of fallacy theory is to fully concede the negativity problem. Fallacy theory, taught exclusively, yields sharks, not arguers. It is a common phenomenon, when teaching a survey of informal logic, to have students ask whether there are any good arguments. One turns them loose on their families over Thanksgiving only to hear, upon their return, that all they could do is call *ad hominem* with their politics-talking uncles. And thus we have another meta-argumentative fallacy, the Eager-Beaver fallacy. A useful analogy to it is that of the college sophomore who takes a course in PSYCH 101 and thereupon diagnoses every person in the dorm with OCD, schizophrenia, or depression. Once you learn a big idea, you can't help but see it everywhere and think that it's a hammer for a world with problems that are nothing but nails in need of pounding. Hundleby correctly observes that informal logic is overwhelmingly taught as *only* the taxonomy of fallacies, and this is precisely the problem. As Hundleby also argues, this is not a reason to *reject* fallacy theory, but a reason to revise our conception of it and to reform the way we teach it (Hundleby 2010: 303).[11]

There are two parts to the *modest* reply to the negativity problem: (i) the mutuality thesis, and (ii) the intrinsic adversariality thesis. Call defenses without intrinsic adversariality, but with mutuality, *minimal* replies. *Modest* replies require both—so if our case for (ii) fails, the minimal defense may still stand with the mutuality thesis. The mutuality thesis is that vocabularies of negative assessment are both part of normative vocabularies and important to their development; insofar as there is *"ought," there are correlate "ought-nots"* that clarify and provide application. The intrinsic adversariality thesis is that a minimal degree of adversariality is part of any argumentative exchange; so as a consequence, negativity is an inescapable component of argument, and any proper theory of argument must be poised for its proper management, not elimination. We will argue first for mutuality, then for the more controversial intrinsic adversariality thesis.

The argument for mutuality begins with what we take to be a simple truth about normative practices—all normative practices have possible meta-languages formulable about them. For example, natural languages have grammars, but the language of grammar need only be *possible*. The same, we think, goes for logic. We have reasoning and arguments, but logic (formal and informal) is a meta-language that makes the rules of the first-order practice explicit. So, the meta-language of logic is a repository of all the rules we (ought to) follow when we reason. All normative practices have the possibility of *error* in their performance, because being bound by rules doesn't guarantee that they are followed. In the case of grammar, common errors are called, for example, *run-on sentences, failures of parallel, subject-verb non agreement*, and so on. The same goes for fallacy theory—common error types are theorized and given names. The point of many of these meta-languages is not only to make the norms explicit and have a theory of their systematicity (and perhaps also a systematic view of the errors, too, as we see in fallacy taxonomies) but to facilitate function of the first-order normative practice. That is, with both grammar and logic, the point is to make the rules (and errors) explicit not just for their own sake but for the sake of self-conscious and reflective normative practice. When a normative practice is self-consciously assessed, the variety of errors surveyed clarifies the norms, and the newly clarified norms allow practitioners to refine their first-order practices and also to find errors that had previously escaped their critical gaze from before. And so with a meta-language, particularly the meta-language of criticism, normative practices evolve as the kinds of practices we can self-reflectively endorse. Fallacy theory, then, is part of a larger dialectic of rationality unfolding, the norms of reason exfoliating against where we make errors in trying to follow and enforce them. And so we see, for example, with fallacies, the importance of their belonging to a language of argument and argument assessment, since they allow not only for criticism of particular arguments but also for assessments of well-run dialectical exchange as being free of particular destructive fallacy forms and for the puzzlement of many when bad arguments occur that don't yield to easy classification.

We can press the grammatical analogy a bit further. Fallacy pedagogy, as distinct from fallacy theory, is aimed at adults. Critical thinking, informal logic, introduction to logic are courses mainly taught at universities, and fallacy handbooks, websites, and various YouTube videos are aimed at grown-ups. Grown-ups might have what we call the *Peirce problem*. In his essay, "Fixation of Belief" Charles Sanders Peirce tellingly observed:

> Few persons care to study logic, because everybody conceives himself to be proficient enough in the art of reasoning already. But I observe that this satisfaction is limited to one's own ratiocination, and does not extend to that of other men.
>
> (1877: CP 5.223)

Adults and even college students come to the study of informal reasoning already, as it were, fully formed. They aren't learning now to make arguments. They already know how to generalize about cats, or how to make inferences about Finland's gross national product. What they need is some fine-tuning on what they're already doing passably well. Indeed, if learning to reason by fallacy were a way of teaching argument for the first time, it would be a manifestly terrible idea. Imagine learning a second language by learning a series of mistakes. However, since our students are fully formed, and often fully confident, a constructive approach to argument is wasted on them. For the same reason, teaching grammar to adults does not involve a return to how to construct grammatically appropriate sentences. Rather, adults improve their grammar by correction. Thus the language of split infinitives, run-on sentences, dangling participles, and comma splices is developed for calling attention to common grammatical shortcomings for folks who otherwise do just fine but need to polish up where things can get dicey or challenging. Well, the same goes for logic classes, and the fallacies particularly.

But there's a further challenge adult learners pose, as Peirce points out: they are often quite satisfied with their proficiency at reasoning. For this reason alone such courses are often *required*. And often, the first step in such courses is that the learners need to be convinced that there is something to learn. A clown parade of mistakes is just the thing. However, as what we've called the Anna Karenina problem for fallacy theory has it, happy families are all alike, but unhappy ones are unhappy all in their own way. There is one way to get it right, and myriad ways to get it wrong.

In this regard, consider the mutuality thesis to be an explicit version of the earlier modest replies to the generality and scope problems—fallacy theory amounts to the development of a critical meta-language that is dialectically heuristic in its first-order application, but is in the service of broader norm clarification in its systematic articulation. The process is open-ended, because the phenomenon explained and discovered is a moving target—how we argue is, in many ways, influenced not only by what we are reasoning about, why we are reasoning, but also in terms of *how we critically talk about our reasoning.* The norms, then, will and ought to evolve as we develop norms of criticism. Negativity, then, is a necessary component of self-conscious reasoning. Keeping track of how things can go wrong is part of our commitment to what's right—that's how fallible creatures do their reflective best. But this is not to say that the negativity of argument should be our sole focus. In fact, the negative critical components of fallacy theory need to be integrated with other programs and objectives, such as argument repair, clarification of important but unacknowledged norms, and even establishing understanding between deeply opposing views. Finally, it is important that those working on fallacies and teaching them not to let the emphasis on this critical edge of argumentative normativity become their sole pedagogical focus. In fact, it seems incumbent on those working in argumentation theory to write more textbooks integrating the fallacy-approach to argument assessment with the broader objectives of informal logic and argumentation. Too many textbooks on informal logic and critical thinking are written by people with no detectable knowledge of the broader reach of argumentation theory beyond the fallacy systematics.[12]

The intrinsic adversariality thesis is considerably more controversial than the mutuality thesis. Both Hundleby (2009, 2010, 2013) and Rooney (2010, 2012) have argued that adversariality may be a dominant paradigm, but it is both a bad one and an optional one. Consequently, it should be foregone. More cooperative communicative models are available, so intrinsic adversariality is, on this view, indefensible. Further, it is clear that the presence of adversariality in argumentative exchange can subvert the broader epistemic objectives of argument (that of pursuing the truth and development of understanding), as it is clear that many will forego argument's adversarial program who are not comfortable with it, adversariality puts many disadvantaged groups at further disadvantage, and it creates retrenchment in the face of further criticism. Plus, relevant to the concerns of this book, we are tempted to straw man our opponents.

Hundleby and Rooney's arguments target primarily Trudy Govier's case for minimal adversariality, showing that, as Hundleby puts it, "we may exchange reasons without opposing one another's ideas—never mind opposing one another personally" (2013: 239). This point is correct, and we think it scores the right critical challenge on Govier's model. Govier's model proceeds from the premise that if one's audience must be on the receiving end of an argument, one must presume that they need correction. As Govier frames it:

> Those who hold not-X are, with regard to the correctness of X and my argument for X, my opponents.
>
> (1999: 244)

Again, our modest strategy is to concede Hundleby and Rooney's point, but make the case for a revised notion of minimal adversariality. We've made the case for the intrinsic adversariality of argument elsewhere (Aikin 2011a, 2017, forthcoming; Aikin and Alsip Vollbrecht 2020; Casey 2020a). But here's a brief outline of the view and our case for it.

There are two distinct cases for the view that argument is intrinsically adversarial: a dialectical case and a doxastic case. For the dialectical case, we argue when we recognize a controversy, when the view we think needs rational support is either actually rejected or has some reason that must be given for it over some relevant alternative. In either case, what need to be given are reasons that play a *contrastive role* between the target proposition for support and some set of alternatives. That is, reasons play the role they do in argument because they favor some thesis over a class of others. So, the argument that "John should ride his bike to work because he could use the exercise" is good if the alternatives are taking the bus or driving, but if walking is a reasonable alternative, too, then the argument's not got quite the shine it had before. Reasons are reasons and have their bite in favor of what they are in favor of only against a contrast of alternatives. In light of this contrastive view of reasons (which we will develop in Chapter 6), arguments always have dialectical complexes, where one's reasons given have the support or critical force they do only if they show that the target proposition is favored or not over a set of contrasting alternatives. In this regard, then, argument is intrinsically adversarial. For arguments to work as reasons supporting a thesis, they must do so by also showing that the competing theses are not favored by those reasons. This is the core of the controversiality conditions for argument that Govier had so famously identified, but it does not require there be actual opponents holding those alternative views. Rather, it only requires that those alternatives be intelligible and deemed relevant considerations under the circumstances. When one argues, even when one collaborates with another to construct an argument for a shared view, there is an opponent that must be represented as at least the contrasting alternative to the target thesis as a consideration that must be addressed.

Notice that the adversariality we see as intrinsic to this structural feature of argument is an adversariality that does not have to take the form of aggression, dismissiveness, or antipathy. While reasons can be adversaries because of dialectical contrast, they can't be aggressive, dismissive, or hostile. This is a feature of the humans who have the reasons. But it is with humans that we (usually) argue, so some account must be given of the adversariality of human interaction. For it is clear that it is not nothing. This brings us to the doxastic case for intrinsic adversariality.

Let's begin by noting that to argue with someone means in some sense to interact with their beliefs. And there are two features of beliefs that bear on the problem of adversarial argument: the causal nature of belief and the consequences of belief on a person's autonomy. With regard to the first, on most accounts, we do not hold our beliefs voluntarily. Try as we might, we cannot will ourselves to believe some particular proposition. We couldn't pay you enough to believe that the current Pope is a Baptist or that the number of stars is prime (presuming you don't believe these things now). This is the thesis of doxastic involuntarism. Now it clearly does not mean that we are passive victims with regard to our beliefs. We are not. We can, for instance, do a number of things to regulate our beliefs. The main thing we do to regulate them is *engage in arguments*, that is, *listen to reasons*; or, what is almost the same thing, *not engage in arguments*. We listen to them in the hopes that the good ones will take root in us and the bad ones will be revealed as bad and so not have any effect. But the only way to achieve those results is more argument—that one argument was good and so should be believed, the others are bad and so should be rejected. But the fact that we have partial control renders the adversarial problem all the more pressing, not less. For our argumentative strategies, including straw manning, are often designed specifically to exploit our partial doxastic control. More importantly, the partial control we have over our own beliefs is counterposed with the full control our interlocutors may have in proposing evidence for our belief. That makes us vulnerable to their work on us with their reasons and proposals. Try as you like, you can't help but think of a chubby chipmunk, once we've proposed you do.

Now, to turn to the second feature of the doxastic case, our beliefs are central to our identity as persons. Changing them—or, importantly, maintaining them—can be a matter of profound significance for us. This is especially true given the partial control we have over the event of the changing, or, crucially, the maintaining of our beliefs. We are, in other words, victims in a sense of those who help us bolster our long-standing prejudices as we are of those who endeavor to convert us to a new position. This underscores the deep moral significance of the exchange of reasons. It also addresses a basic presupposition in the case against adversarial argument: that the fact of arguing is what brings the adversariality. Here we have argued that it is the nature of reasons, for the dialectical case, and the nature of beliefs, for the doxastic case. An important conclusion of the intrinsicist thesis is that argument, and even the absence of disagreement, is adversarial already.

If this line of reasoning is correct, the crucial element to training in fallacy theory is to mitigate the escalation of adversariality from, on the one hand, the useful and productive sources of critical feedback for argument to work to, on the other, the exchange of insults. So long as the cooperative exchange is critical discussion, dialectically minimal adversariality need not be any impediment to arguments given with good will.[13] Moreover, notice that there can be collaborative elements that emerge from these considerations such that one can truly value critical questions and challenge, not as personal attacks or even rejections of one's point of view but as the kind of useful resistance needed to craft the case for any controversial view.

1.2.5 Taking Stock

Fallacy theory, properly framed, is a domain with contested target phenomena, and as a consequence, contested applicability and normativity. This comes as no surprise to anyone familiar with philosophy's long history of fits and starts on a variety of issues. And so it goes for the domain of fallacy theory. The generality and scope problems are representative of a requirement that domains of research be well-ordered and (at least potentially) finite, and these are reasonable expectations for many areas. But the phenomenon of reasoning is a moving target, as our vocabularies of evaluation change the phenomenon. In fact, the important thing is that they *do* change the phenomena. That people now invoke *ad hominem* attacks or slippery slopes in the midst of arguments is testament to the contribution fallacy theory makes to self-conscious argumentation. Further, the negativity of this critical vocabulary is itself intrinsic to this program of bringing normative practices to self-consciousness. The terms of critique are both part of the first-order practice of argument (as the intrinsic minimal dialectical adversariality argument runs) and part of our grasp of that practice as rule-bound (as the mutuality argument runs). And it is a tool of intellectual self-defense (as the doxastic adversariality argument runs). The conclusion, then, is that fallacy theory is messy and adversarial, and necessarily so. What's required of us, then, as argumentation theorists, is not that we reject fallacy theory or reform it to the point of being non adversarial but that we develop research and teaching programs that (a) maintain a minimum of well-orderedness to research and (b) mitigate the potentiality of adversarial escalation in argument. So programs of *argument repair* alongside fallacy identification must be taught, and we must keep track of the way our critical vocabulary returns to and influences the practices it is designed to describe.

2

Fallacy Names and the History of the Straw Man

This book is about a fallacy with a rather poetic name—the *straw* man. This name, and its metaphorical imagery, has itself inspired several other metaphorical names: the *weak* man, the *hollow* man, and now the *iron* man and *burning* man, all extending the poetic representation of the arguments and opponents as made of straw. We will turn to those new names in the chapters that follow. For now, we will attend to the history of the straw man and to the general program of naming fallacies. Naturally, the metaphor of an opponent made of straw brings along with it a host of associated notions, some of them problematic. And, like many fallacy names, it has an interesting, and sometimes surprising, history, one that sheds light and uncovers insights into fallacy theory and the proliferation of new fallacy names and concepts. But this, as we said, doesn't come without problems and assumptions. Let's start there with fallacy names, then, because *straw man* seems like a perfect fit.

2.1 The Problem of Fallacy Names

While few people, and we're including philosophers in this, are familiar with argumentation theory and its sub-discipline fallacy theory, many are quite conversant with fallacy names—*ad hominem*, begging the question, *post hoc ergo propter hoc*, *secundum quid*, no-true-Scotsman, the Texas Sharpshooter fallacy, the straw man (of course, but probably not the iron man—yet) and the myriad hundreds or perhaps thousands of others. The names—perhaps on account of their poetic mnemonics or esoteric appeal—do a lot of the argument theoretical work. This is unavoidable, because, of course, names of fallacies are regulative terms; but it's unfortunate, because the names are actually kind of a mishmash that's really poorly suited to the work it's been asked to do. Let's take a quick tour.

The most notable, and most deceptive, category of fallacy names are Latin names. It's clear that the Latin of it all carries a lot of weight—as if to say that something in Latin has survived the Middle Ages, at least, and so is worth listening to. Things stated in Latin have *gravitas*. In addition to that, this presentational feature has what we've called in Chapter 1 the *Harry Potter* effect; its mere invocation is equivalent to casting a spell of authority over a wayward interlocutor. But, like the Latin in the Harry Potter novels, it's not always quite what it seems. While some of the Latin names for the fallacies are translations of Aristotle's Greek terms (e.g., *petitio principii, ignoratio elenchi*, and even *ad hominem*) some others, such as *ad verecundiam*, and *ad ignorantiam*, find their origin, strangely, in Locke's (1689) *Essay Concerning Human Understanding*, which was written in English. What is even odder about Locke's invention of these terms is that they were not at all described as *fallacies* but rather as common forms of argument. *Ad hominem*, for example, means *at* or *with regard to the person*, and is a kind of argument where one presses "a Man with Consequences drawn from his own Principles, or Concessions" (1975: 686) thus *ad*, or with respect to, some *hominem*. It's an argument directed at a specific person's views. This is far from the sense of the *ad hominem* as an attack on the person and not the argument.[1] When Locke wrote his *Essay*, fallacy theory, and its laundry-list of names of fallacies, was in its infancy (and of little interest to him). The medieval treatments of fallacies that preceded him were largely content with discussions, sometimes quite detailed, of Aristotle's original 13, as noted by Hamblin (1970). The real codification of the *ad-* arguments as fallacy names begins with Isaac Watts (1825). Jeremy Bentham, or his ghostwriter "Peregrine" in his *Book of Fallacies* (1824), sums up the case for using Latin names:

> To the several classes of fallacies marked out by this subdivision, a Latin affix, expressive of the faculty or affection aimed at, was given; not surely for ostentation, for of the very humblest sort would such ostentation be, but for *prominence, impressiveness, and thence for clearness*.
>
> (emphasis added; pp. 9–10).

For what it's worth, Bentham has a theory of the Latin name: it's not a name of the fallacy (he gives those in English, often hilariously) but a name of the fallacy's *target*. And to the scattered Latin names he adds several of others, including the *argumentum ad odium* (appeal to hate), *ad metum* (fear), *ad quietam* (quiet), *ad socordiam* (slowness), among others. Some of these sprout several different English fallacy names, the fallacy of no-consolation and the procrastinator's fallacy among them.

Another class is English or Modern Language *descriptions* (to be honest, we're unaware of names for fallacies in languages other than English or Latin). There are two basic kinds of descriptions. First, descriptions of the *failing or bad doing*: what's wrong is that you generalized *hastily* or you *affirmed* the consequent, or you inferred *from ignorance*, or you *begged the question*, etc. Oftentimes these are Latinized (*argumentum ex silentio* and *petitio principii*). A second type is a description of the thing mismanaged but *not* the mismanagement: *sunk cost* or *base rate* these are failures regarding these things, thus the *free speech fallacy*. A third group includes partially descriptive names that are intended to be memorable. These are not descriptions but rather descriptive *analogies* or *metaphors* for the failure in question. The straw man (and all of its various sub-types) belongs to this class. So does the red herring, the smoke screen, and poisoning the well. Finally we have fallacies named after people (who invented them or whom they regard). These are not descriptive at all. Some famous examples: Godwin's Law, Poe's Law, and the Gish Gallop, or properly Harry-Pottered, *ad Hitlerum, Lex Poensis,* and *Cursus Gishis*. Obviously, Latinizing them makes them extra funny—not something Bentham noticed.

2.1.1 The Foul Names Problem

Now that we're clear on the nature and the rough history and typology of fallacy names, we'd like to consider some (sometimes) unfortunate problems. The first problem is that the mere existence of fallacy names, as noted in Chapter 1, casts an adversarial shadow over argumentation. Consider that a common analogy among teachers is that a logical fallacy is a kind of argument foul. So, as tripping is a kind of soccer (football) foul, *attacking the person* is a kind of argument foul. We've used this analogy ourselves. The analogy works if we view both activities (arguing and soccer) to be rule-bound and competitive activities. If we're going to have a debate, then we not only play to win but there are rules to follow (stick to the subject, don't straw man my position, etc.). So, just like if we're going to play soccer, we're going to play soccer, so don't trip me. But the analogy has its limits.

In the first place, the playing of sports, barring some post-apocalyptic scenario (fingers crossed!), is optional. To the extent that it's optional, the rules are conventional and as a result agreed upon (and also arbitrary). You don't have to play, but insofar as you do play, these are the rules. Argument is also optional, but only some of the time. How much of the time is an open question. There are times it is not optional (in court, say, or at work, where you are called upon to give reasons for something or defend your actions or position), but the reasoning, in a way, behind it all isn't optional.

The rules of sports are also almost completely unlike the rules of argument in their general purpose. The rules of sports have two aims. First, they *optimize competition*. They are designed to make the game *fun* and *challenging*. The game is no fun if the rules allow one team to dominate or for the it to be over too quickly. (Baseball games, for instance, *never end,* and there are discussions of rule-changes for shortening them with an eye to mitigating fan misery but also reducing hot dog sales.) A second feature of the rules in sports is to make sure people don't get hurt by overzealous competition. Thus there are prohibitions against potentially harmful activity (like face-masking or tripping). This we guess also makes it fun, or at least so that it can be fun, given that you're not worried about those harms.

This description of sports has its limits as well. For, too often people take violations of the rules to be a part of the game. If a player has you beat in American football, you take a penalty to stop them rather than suffer the certainty of their scoring. Thus the rules of the sport quickly divide into essential rules and inessential rules or violable rules. An intuition, for what it's worth, of many of our students is that there's a deeper set of rules determining when you can violate the first set of rules. These rules might be the principles of fair play or something like that. The analogy, so it appears, faces some challenges. We'll return to these in Chapter 7.

2.1.2 The Harry Potter Problem

Aside from potentially problematic (or just inappropriate) adversarial notions, fallacy names, unlike fouls in sports, can make argument less worthwhile and so less productive. One frequent but troublesome practice is invoking fallacy names in place of substantive discussion. As we described it in Chapter 1 and here above, *the Harry Potter fallacy,* the tendency to invoke fallacy names like so many magic spells is especially tempting when the names are in Latin. Calling *argumentum ad verecundiam* on someone's invocation of a sketchy authority is rather satisfying, we must confess. We've already seen that the standard pedagogical methodology very much favors this practice. Exercise sets in textbooks are so many victims lined up for execution. Sadly, the pedagogical lesson often seems to end here: this is what the point of it all is, it seems. So, armed with their new vocabulary, students take special relish in throwing around the names. It's doubly so if the target doesn't know what the phrase *argumentum ad verecundiam* means.

A second adversariality problem lies with the names themselves. That is a pathology of meta-languages, call it *weaponized metalanguage* (as in Aikin and

Talisse 2020). Sadly, as we've conceded in Chapter 1, there appears to be no obvious escape from this problem. As one seeks a new regulatory ideal, that ideal becomes a part of the game itself, subject to the same old or even new abuses. The history of fallacy names is an interesting case in point. For, not all of the fallacy names are weaponizable to the same extent and a brief tour of the various names reveals an interesting mishmash of approaches to argumentation. Further, aside from the employment and the weaponization problem, some of the fallacy names themselves seem to embody problematic adversarial notions. Though it wasn't originally intended this way, the *ad hominem* is *attack the person* and, of course, the *straw man*, in its very first manifestation, implies an attack on straw woven into human form. In fact, this is precisely what the metaphor of the man of straw invokes—an opponent who cannot fight back and who is easily destroyed.

2.1.3 The Proliferation Problem

With regard to proliferation, as fallacy theory advances, as new journal articles (and, er, books) are produced, new names enter the lexicon, inundating what some consider to be an already overly saturated catalog. But you don't really need fallacy theorists to fatten up the lexicon; the internet has conferred this opportunity on everyone—so *motte and bailey*, *whataboutism*, and *Godwin's law*. Thus, invoking the famous Tolstoy line about happy families all looking alike and unhappy families being unhappy in their own unique way, we think there is what we called in Chapter 1 an Anna Karenina problem for fallacy theory—unhappy arguments are all unhappy in their own unique way. (Whether good arguments are all good in the same way, we will not weigh in about.) Since fallacy names are so many names of infelicities, in won't be long before we really appreciate the depth of Tolstoy's observation.

Classically, Aristotle identified thirteen fallacies. A recent publication *Bad Arguments: 100 of the most Important fallacies in Western Philosophy* (full disclosure: we contributed two entries) listed (perplexingly) more than 100 of them. But that's only the beginning. There are hundreds more on some websites. It's not the only job of fallacy theory however to invent and popularize new names, that's more like moonlighting. What fallacy theory does is examine and classify types of argument pathology, to use a timely medical metaphor, and diagnose the root causes of disease and, hopefully, discover a cure, or, at the very least, an inoculation (Boudry 2017). But you can't (probably) inoculate yourself against every disease. For that, you will need a healthy immune system.

We addressed a broader version of the Proliferation problem in Chapter 1 as the Scope problem for fallacy theory—that the bounds for the domain seem to be too penunmbral. And so, it seems fallacy theory has no clear demarcational criterion that makes it different from culture criticism, political epistemology, or even journalism. It certainly begins to look too much like a collection of philosophers' pet peeves about their colleagues' argumentative habits. And, though interesting as an expressive enterprise, it's not clear how it's a domain of research. We agreed that the problem is that the variety of ways for argument to fail far exceeds our taxonomies of general types. And even the ones fitting in the taxonomies are often bad fits. In fact, this will be a lesson of our brief tour of the history of the straw man fallacy. So, *non sequitur* and *ignoratio elenchi* are Latiny names for arguments that just don't work, but those classes have little more to say about why they don't work than that they just don't. Our project is to open up unruly boxes of fallacies to see patterns, similarities, and new and unappreciated features. The straw man fallacy itself is the result of this expansion by logicians and philosophers that preceded us, and we see ourselves continuing this project with focused attention on this particular fallacy. And we will note that once we add to the list of fallacy types, these categories for correction can themselves be looped back around and be part of the further polluting of argument. That's the Owl of Minerva problem again, but it's worth being aware of it, at least. That's how we manage the problem and its consequences. Consider again Tolstoy's *Anna Karenina* observation about unhappy families now complicated with the injection of family therapy—the categories from family therapy can contribute to healing an unhappy family and putting them on the road to being happy, or those categories themselves can be yet new factors that contribute to the unhappiness of the family. Any family that needs therapy risks the likelihood that the vocabulary of therapy will escalate the situation that necessitated the therapy. Well, the same goes for fallacy theory. So much of argument is not just reasoning about what we reason about but reasoning about how we are reasoning. Meta-argument is part of argument, and knowing the fallacy list doesn't just improve how we reason about how we reason, it can change how we reason on the first order. That's the high hope of teaching the fallacies and developing finer-grained fallacy theory—that we aren't just getting a better view of argumentative practice, but that the view improves the practice. But the *Harry Potter* fallacy shows that knowing that vocabulary can stand in the way of both the meta-argument and the first-order argument. And, as we'll show in Chapter 6, knowing the fallacy list makes it possible for an entirely new kind of straw manning to be possible, one that Aristotle or his later commentators would not have had on their lists.

Because, in a sense, you would have had to have read Aristotle and some other critical thinking books to commit this specific kind of fallacy of straw man. That is, it's a fallacy that comes out of our skills of correcting fallacies and an audience that's familiar with that practice. Our concepts of good and bad argument can have salutary or pathological looping effects, and the history of these concepts (and the straw man, in particular) shows this. Our objective, then, is to trace the history of how the error of straw manning was identified, how the image of the "straw man" developed and functioned as a heuristic for argument-criticism, and then was deployed as a diagnostic tool of evaluation and correction.

2.2 History of the Straw Man Fallacy

In the wake of this brief but deflationary account of fallacy names, we now turn to a discussion of the straw man in particular. Our aims are twofold. First, we want to trace the history of the straw man fallacy and the straw man name (these happen to be different). Second, we will discuss the straw man as it appears in recent textbooks. We will discuss theoretical aspects of straw manning later in Chapters 5, 6, and 7.

It is surprising that the idea of misrepresenting an interlocutor's argument so as better to refute it, which seems like if anything the most obvious nefarious move to make in an argument, doesn't appear to have been described and named until relatively recently. It is also curious that the straw man or its distant cousins are not included in Hamblin's *Fallacies* (1970), a recounting in often exquisite detail of the undulating fortunes of fallacies and fallacy theory from the time of Aristotle, through later Hellenism, the Middle Ages, and on to the modern period.[2] It doesn't show up there as a part of what Hamblin calls the "standard treatment" and it didn't merit even a passing mention in the index. This is a curious oversight that we intend to remedy here, even though a comprehensive history of the straw man is beyond the scope of this project. As we'll see below, the straw man seems to have grown out of an interpretation, or rather a misinterpretation, of the fallacy of the ignorance of refutation, or the *ignoratio elenchi* (ἐλέγχου ἄγνοιαν) in Aristotle's *On Sophistical Refutations*. Its fortune as a fallacy is therefore bound up, in ways that are both strange and enlightening, with the *ignoratio elenchi*. In the end, it's a telling example of what we think is a persistent *ad hoc* nature of fallacy theory.

It is worth remembering that Aristotle's *Sophistical Refutations* is not really a handbook of fallacies as one finds so many exemplars of nowadays. It's not, in other words, *thirteen fallacies all good reasoners must avoid*. The scope is rather more narrow; it deals with a very peculiar (and still somewhat mysterious) form of question-and-answer adversarial disputation. Roughly, it was a highly formalized version of a Platonic dialogue.[3] The purpose of the text is also at odds with the contemporary handbook approach: Aristotle's interest is not the discursive hygiene of such disputes but rather in how to *commit fallacies* in the course of them and thereby outmaneuver one's opponents (165a19). To this end, he also includes a variety of other argumentative stratagems such as burying your point in a rather long discourse or talking really fast (174a16). Think of this as what we've called the puzzle of effectiveness, only in reverse: *how do I get away with these things?* As a historical matter, those who worry that there is an adversarial edge to the entire fallacy story are not therefore entirely wrong. But what Aristotle meant and how he was interpreted are two different things. And, with regard to the *ignoratio elenchi*, all that is really left of it is the name.

Turning now to the origins of the straw man, one has to squint really hard to see its first traces in the touchstone passage for our genealogy in the *Sophistical Refutations*. But, as we'll see presently, this text was according to later interpreters the basis of their versions of the straw man, so it's worth stopping over for a moment. Aristotle writes:

> Other fallacies occur because the terms 'proof' or 'refutation' have not been defined, and because something is left out in their definition. For to refute is to contradict one and the same attribute-not merely the name, but the reality- and a name that is not merely synonymous but the same name-and to confute it from the propositions granted, necessarily, without including in the reckoning the original point to be proved, in the same respect and relation and manner and time in which it was asserted. A 'false assertion' about anything has to be defined in the same way. Some people, however, omit some one of the said conditions and give a merely apparent refutation, showing (e.g.) that the same thing is both double and not double: for two is double of one, but not double of three. Or, it may be, they show that it is both double and not double of the same thing, but not that it is so in the same respect: for it is double in length but not double in breadth. Or, it may be, they show it to be both double and not double of the same thing and in the same respect and manner, but not that it is so at the same time: and therefore their refutation is merely apparent. One might, with some violence, bring this fallacy into the group of fallacies dependent on language as well.
>
> (NE 5, 167a21–36)

There are a few shades to the *ignoratio elenchi* as Aristotle describes it, and one of them appears (again, if you squint) to be a representational straw man.

One misrepresents the point at issue and gives a refutation of something else, the straw man-style misrepresented thing, so as to shift the focus. To refute a point, after all, means to refute the point in question, not something apparently like it. One way they might achieve this apparent refutation is by, as Aristotle notes, *omitting* some feature of the case. Perhaps it might be an important qualification or a particularly powerful piece of evidence the target of straw manning introduces. This captures one of the core notions of the straw man and one of the core notions of rationality we're discussing here. But Aristotle's actual point here differs hugely from what gets made of it almost 2,000 years later. The *ignoratio elenchi* is a matter of failing at understanding the concept of refutation rather than drawing an irrelevant conclusion (or *non sequitur,* as it often appears in the textbook analysis nowadays). In fact, the *ignoratio elenchi* can be taken in two basic ways, according to Aristotle. In one sense, every one of the other fallacies is an ignorance of what an elenchus is; in other sense, it is a specific failing. Broadly speaking, all the fallacies are really instances of the *ignoratio*—you just are misunderstanding how refutation and argument work. And this is one reason why *ignoratio* is a useful bin into which one can toss fallacy types that don't fit into one's running typology. In the more narrow sense, though, it's ignorance of how *refutation* of a particular dialogical opponent works. There are the roots of the straw man as it is theorized for our purposes.

The fortune of Aristotle's logic in the Middle Ages is a tale in itself, one very far beyond our interests here.[4] But we would like to mention, so as not to leave a huge gap in our narrative, that later medieval interpretations of the *ignoratio elenchi* hew fairly closely to Aristotle's sense of it and the reveal little or nothing of what was to become of it a few centuries on.[5] As far as we can tell, the tradition of broadening Aristotle's *ignoratio elenchi* to cover cases of dialectical misrepresentation begins with Antoine Arnauld (1611–94) and Pierre Nicole (1625–95). In *Logic, or the Art of Thinking* (1662), they discuss fallacies, or "sophisms," in two short sections at the end of section III (On Reasoning). They divide fallacies into two groups: those that occur in science and those in popular discourse. The first of the fallacies that occur in science is the straw man *ante litteram*.

> This sophism is called by Aristotle *ignoratio elenchi*, that is to say, the ignorance of that which ought to be proved against an adversary. It is a very common vice in the controversies of men. We dispute with warmth, and often without understanding one another. Passion, or bad faith, leads us to attribute to our adversary that which is very far from his meaning, in order to carry on the contest with greater advantage; or to impute to him consequences which we

imagine may be derived from his doctrine, although he disavows and denies them. All this may be reduced to this first kind of sophism, which an honest and good man ought to avoid above all things.

(p. 242)

We might now have an answer, by the way, for why the straw man does not have one of the famous Latin names: it does. It's the *ignoratio elenchi*. Arnauld and Nicole seem to have reflected upon the *ignoratio* and then contemplated other ways one can be ignorant in the context of a dispute. So far as it goes, this is a fairly serviceable description of the straw man as one might encounter it these days; it suggests that straw manning is a creature of adversarial arguing, where the stress of contest and the desire to emerge victorious make us poor listeners. But the inclusion of imputed consequences of a view suggests that there is still a lot of room for clarification. For the question would be, in the case of consequences, whether they follow logically from someone's view, not whether they hold them; so, if anything, the fallacy of that clause would be something else. This early version of the straw man, as they note, is a dialectical, not a logical, failure. Worth mentioning, if only for a few laughs, is that Arnauld and Nicole go on to accuse Aristotle of being one of this fallacy's most prominent practitioners.

> It could have been wished that Aristotle, who has taken pains to point out to us this defect, had been more careful to avoid it; for it must be confessed that he has not combated honestly many of the ancient philosophers in reporting their opinions.
>
> (242)

Anyone with any sympathy for one of the Presocratics under Aristotle's withering and dismissive gaze in the *Metaphysics* will feel a little rush of appropriate elation at this bit of philosophical comeuppance. Arnauld and Nicole add detail to the accusation, but what is interesting is the shift in context not noticed by them or really anyone in their wake. To repeat, Aristotle's *Sophistical Refutations* is a rather different text from Arnauld and Nicole's *Art of Thinking*. Aristotle's is a guide for using fallacies and other trollish tactics in the context of a highly formalized adversarial debate. But Arnauld and Nicole are unaware of this, and it's as if they read the *Sophistical Refutations* like the *Prior* and *Posterior Analytics*, texts where Aristotle outlines some of the basics of deductive argument and then philosophical method. They read it as an outline of something normative. And this underscores the fact that Aristotle, as perhaps evidenced by his own practice, did not have much to say about ordinary philosophical disputes other than to diagnose them as being more confusions of what was obvious. Though, of course, he ought perhaps have been a greater friend of truth here.

Fast-forwarding about a century and traversing the English Channel, we come to the first actual employment of the term *straw* in a work on logic. Isaac Watts (1674–1748), in *Logick; or, The Right Use of Reason* (1725) extends the interpretation of the *Ignoratio Elenchi* to the straw man, and even employs the idea of a straw image one attacks in his description. His sense of the *ignoratio elenchi* follows the general view that almost any sort of broad irrelevance counts as an instance. He also couches his account in the same language as Arnauld and Nicole:

> Disputers when they grow warm, are ready to run into this fallacy; they dress up the opinion of their adversary as they please, and ascribe sentiments to him which he doth not acknowledge; and when they have with a great deal of pomp *attacked* and *confounded these images of straw of their own making*, they triumph over their adversary as though they had utterly confuted his opinion.
>
> (270, emphasis added)

Watts's image of straw is a kind of effigy and the stress seems to be on its emptiness rather than, as we'll see in a moment, the ease with which it is knocked down. Watts has in any case a better account, so it appears to us at least, of how this fallacy emerges from the *ignoratio elenchi*. Well, it's better but not perfect. The general idea seems to be that the *ignoratio elenchi* is a general failure to refute a point. One method of failing is logical (as we saw above with Arnauld and Nicole), one method is dialectical, where you misrepresent the opponent's expressed content under some identifiable context. Watts adds further detail to the dialectical version of the *ignoratio elenchi*, noting that it's an attempt to score easy points.

> It is a fallacy of the same kind which a disputant is guilty of, when he finds that his adversary is too hard for him, and that he cannot fairly prove the question first proposed; he then with slyness and subtlety turns the discourse aside to some other kindred point which he can prove, and exults in that new argument wherein his opponent never contradicted him.
>
> (270)

It is still early days for modern fallacy theory, so we've yet to encounter puzzles, such as the puzzle of effectiveness mentioned in the Introduction (and discussed later in Chapter 5). And the advice is similarly basic:

> The way to prevent this fallacy is by keeping the eye fixed on the precise point of dispute, and neither wandering from it ourselves, nor suffering our antagonist to wander from it, or substitute anything else in its room.
>
> (1824: 270)

Note that this advice still suggests the face-to-face dispute, and so a *dyadic* relation between arguers, that there are two and only two in the exchange. We think this is manifestly insufficient for the explanatory account of how straw manning works, so we will propose a tri- or polyadic dialogical account later. But part of the explanation for why the account in the dyadic form is so dominant is the genealogical point that all this nascent and growing theory of the straw man traces its history back to and is still inspired by Aristotle's account of how to handle argumentatively dyadic dialogue in *Sophistical Refutations*.

Curiously, the image of the "straw" man as a rhetorical figure does not appear to be an invention of Watts. The phrase was relatively common in Elizabethan England to mean pretty much exactly what many people take it to mean today. English clergyman Thomas Gataker (1574–1654), writing in reply to James Balmford on a question of gambling (which Gataker seems to have favored to some extent), provides an assessment of the latter's argument:

> True, I presume Recreation to be lawfull in generall. Dare Mr. B. or any man denie it? And yet I presume it not without proofe neither. *But you must remember that Mr. B. here sighteth not with me or mine Argument, but with a man of straw of his owne making.* In his Assumption this is presued of Lotterie: whereas I presume it of recreation, not of it.
>
> (1623: 118–19)

The title of this work, we'd like to point out, does not pull any punches regarding the quality of Balmford's arguments: *the imbecillitie of his arguments produced against the same further discouered*. The image, in any case, appears therefore to have been a fairly common one. Jacqueline Broad (2020) speculates that it has its origin in the figure of Jack Straw, who led an unsuccessful peasant revolt in 1381. His image as a straw effigy was included in the pageant procession of the Lord Mayor of London. Broad notes a comment by Mary Astell in the margin of a text by Bayle to this effect: "He [Bayle] sets up a Jack Straw that he may throw stones at him" (Broad 2020). To this day, an effigy of Jack Straw, a kind of scapegoat for the sins of the world, is burned to ring in the New Year in Hungary. This practice partially explains why some nowadays associate straw manning with fire, and we extend it later, in Chapter 6, to aggregative cases with what we call the *burning man*.

Our short history has so far revealed an Aristotelian origin of the fallacy, an explanation for why there is no unique Latin name, and an origin of the English straw man image. Looking forward, beginning with De Morgan, fallacy theory begins to develop significant heft and nuance. Many of the dialectical features central to argument analysis have their origin in nineteenth-century British texts, especially De Morgan, Whately, and Mill.

Richard Whately (1787–1863) has by our lights the most extensive and theoretically nuanced discussion of straw manning in the early literature. We have already noted that he is attuned to the question of the argument form's effectiveness, but his account of the straw man also shows that he's aware of what we've called the meta-argumentative puzzle as well. The straw man, in other words, is what you get when you engage in arguing about argument. Whately's account of the straw man in *Elements of Logic* (1826) grows out of an account of, who would have guessed, the *ignoratio elenchi*. Notably, Whately finds himself to be rather stuck with the name *ignoratio elenchi* on account of its common use, though he would like to set it aside (178). Whately considers the *ignoratio elenchi* to be a "non-logical" or "material" fallacy the decisive feature of which is that the conclusion *does* follow, but it's irrelevant to the discussion at hand. This broad analysis also holds of the *ad hominem, ad verecundiam*, and other "ad" fallacies of relevance. He offers a few separate accounts of the straw man as an instance of the *ignoratio elenchi*. The first consists in the proving "what was not denied" or disproving "what was not asserted" (p. 236). Such practices, he writes, are an "affront" for they "[attribute] to a person opinions, etc., which he perhaps holds in abhorrence" (ibid.). Another version of the straw man can be found in what he calls the "fallacy of objections." Interestingly, the fallacy of objections reverses the dialectical order of the straw man: rather than putting the distortion feature first, it foregrounds the dialectical technique which is a consequence of the distortion. So the "objections" in the fallacy of objections are what are generated by various kinds of straw man distortions. These objections are then leveraged as evidence against a view.

> Similar to this case is that which might be called the *Fallacy of objections i.e.,* showing that *there* are objections against some plan, theory, or system, and thence inferring that it should be rejected; when that which *ought* to have been proved is, that there are *more*, or *stronger* objections, against the receiving than the rejecting of it. This is the main, and almost universal Fallacy of anti-Christians; and is that of which a young Christian should be first and principally warned. They find numerous "objections" against various parts of Scripture; to some of which no satisfactory answer can be given; and the incautious hearer is

apt, while his attention is fixed on these, to forget that there are infinitely more, and stronger objections against the supposition that the Christian Religion is of *human* origin; and that where we cannot answers all objections, we are bound in reason and in candor to adopt the hypothesis which labors under the least.

(Italics in original, 241–2)

There are two straw men in this passage. One of them is what we now call the weak man or the selectional form of the straw man when he says "to some of which no satisfactory answer can be given." He returns to this same thought a few pages later (245). The other straw man is a kind of hollow man in the guise of a kind of empty objection one can quickly and easily defeat. Put another way, the first is that there exist objections against weak parts of a proposal; therefore, the proposal ought to be shelved. The second is about how easily objections are handled. This observation highlights what is innovative about Whately's view is his attention to the broader dialectical objectives of straw manning and to argument's doxastic adversariality (mentioned in Chapter 1). The point, he notices, is not to win some dispute with an opponent (as in the case described above by Arnauld and Nicole) but rather to create the impression in the minds of an onlooking audience that the dialectical terrain does not necessarily favor your opponent. What is even more interesting about this observation is that it relies on the "incautious hearer"—that is, an inattentive audience unable to detect the misrepresentation. Whately is, then, the first to try to address the puzzle of effectiveness by denying the dyad of the argumentative relation and invokes a triad of speaker, target for straw man representation, and an audience for the representation. We think this is of real significance. Whately notices more dialectical malfeasance. The straw man, he argues, is a part of a broader strategy:

> The very same Fallacy [the fallacy of objections] indeed is employed (as has been said) on the other side, by those who are for overthrowing whatever is established as soon as they can prove an objection against it; without considering whether more and weightier objections may not lie against their own schemes; but their opponents have this decided advantage over them, that they can urge with great plausibility, *"we do not call upon you to reject at once whatever is objected to, but merely to suspend your judgment, and not come to a decision as long as there are reasons on both sides:"* now since there always *will* be reasons on both sides, this *non*-decision is practically the very same thing as a *decision in favor of the existing state* of things. "Not to resolve, is to resolve." The *delay* of trial becomes equivalent to an *acquittal*.
>
> (1855: 243–4 Italics, in original)

To be clear, Whately is saying that the mistake (in this case) consists in the superficial considerations of the existence of objections without regard to their quality. It's a way of ignoring the argument and just focusing on the mere existence of arguments. It seems that these are two different fallacies of meta-argumentation. Nonetheless, Whately's point about the dialectical tie is interesting: you distort the dialectical terrain ("there are weighty reasons") and then you call a draw. Calling it a draw then maintains the *status quo ante*. Particular skeptical results are allies to conservatism.

As we will discuss later in Chapters 3 and 4, Whately's observation is that to distort one argument often is a means of exaggerating the strength of another. You strengthen the objections by not covering them or evaluating them but rather by naming them. *There are objections to this view*, you say; *they mean that it's not universally seen to be true* so therefore *it must be weaker*. Time, energy, attention, and so on prevent you from evaluating these reasons, so you call it a draw or convince like-minded people or audience properly primed with antipathy for your target to infer that it's *more complicated* or *more fraught* than it would otherwise appear. So, in the end, you're warping the strength of these positions (first, by exaggerating their strength and then by exaggerating the weakness), and this works to promote an argument to a state of draw. Another important lesson from this discussion of Whately's is that *you don't always need to argue to win*; playing for a tie is sometimes sufficient.

Another consideration here is the exaggeration by the one who employs the fallacy of objections of the criteria for a successful argument. The implication is that a successful argument must be free of all objections. That there are objections *at all* suggests the argument is not as strong as it needs to be to carry the day. On the one hand this ignores the quality of the objections themselves, as Whately notes, and on the other hand, it suggests that success in argumentation is a very high bar. So you move the goalposts in a way—you make them much much smaller so that nothing can really get through them. This also shows that there is another kind of straw manning that isn't directed to specific arguments but rather broader disagreements. One can distort disagreements by implying arguments on one side or the other are stronger than they are. This is not the distortion of a particular argument to defeat it, as in the classical argument-centered straw man. Rather, it's view on multiple and disparate arguments. It's a superficial approach to disagreements and a disengaged approach to arguments in order to draw some unwarranted conclusion about the whole lot of them. It's a broader problem because it does not have a specific victim; there isn't a straw manned person or a single victim but rather views or issues generally. The takeaway is that whole debates can be the targets for straw manning. And notice, finally, that it all happens as a form of *meta-argument*, as the inferences made are not about the things reasoned about, but about the quality of the reasons in light of the fact that there are objections.

Augustus De Morgan in *Formal Logic: Or, The Calculus of Inference, Necessary and Probable* (1847) breaks with the tradition of considering the straw man (which he does not name) a species of *ignoratio elenchi*. Instead, he appends a rather lengthy account of it at the end of his discussion of fallacies. He notes, "It is not uncommon, in disputation, to fall into the fallacy of making out conclusions for others by supplying premises" (281). What he goes on to describe is an interesting variation on straw manning: *the imputation of commitment*. Crucial to this is the idea that people are responsible for commitments that they explicitly accept. The extent to which they are, by the way, is a live topic in logic (and in many ways the heart of the problem of adversarial argumentation).[6] Otherwise, as De Morgan argues, "Any sect of Christians might be made atheists by logical consequence, if it were permitted to join together the premises of different sections among them into one argument." In what has to be a landmark in the journey from Aristotle's initial account of the *ignoratio elenchi*, De Morgan diagnoses the cause of this problem as viewing argument as a competition: "This is a fallacy which, however common, could easily be avoided, and would be, if those who use it cared for anything but victory" (281).

John Stuart Mill is an admirer of Whately's text and he follows it in many things. We won't rehearse all of that here, but Mill approvingly remarks that "The works of controversial writers are seldom free from this fallacy." He then offers a series of examples, including Dr. Johnson's famous refutation of Berkeley's idealism (the view that all things are, in the end, mind). Johnson kicks a stone and announces that he's thereby refuted Berkeley's idealism. If only silly philosophical views had such easy paths to refutation! Hilariously, Mill suggests that the *ignoratio elenchi*, or the straw man, finds itself in expressions, gestures, as well as words, and philosophy is a place for their stark expression. Aside from simply grinning at Berkeley's idealism, one also finds a famous gestural refutation:

> The argument is perhaps *as frequently expressed by gesture as by words*, and one of its commonest forms consists in knocking a stick against the ground. This short and easy confutation overlooks the fact, that in denying matter, Berkeley did not deny any thing to which our senses bear witness, and therefore can not be answered by any appeal to them.
>
> (1882: 1011)

One can hear in this what Aikin and Talisse have called *modus tonens*, the act of repeating an interlocutor's argument back in a funny, oafish, or incredulous voice (2019: 93). More on that anon. Mill is very confident that exemplars of this sort of *ignoratio elenchi* are abundant and easy to detect and, for those reasons, (much to our great dismay) not really delight, by his lights, worth cataloging:

> It would be easy to add a greater number of examples of this fallacy, as well as of the others which I have attempted to characterize. But a more copious exemplification does not seem to be necessary; and the intelligent reader will have little difficulty in adding to the catalogue from his own reading and experience. We shall, therefore, here close our exposition of the general principles of logic, and proceed to the supplementary inquiry which is necessary to complete our design.
>
> (1012)

Perhaps Mill will forgive us if we ignore his advice not to add to the unruly bestiary of such fallacies. He is wary of what we called in Chapter 1 the Scope problem for fallacy theory, but we are not quite so chastened by the challenge of it. In fact, we think that not giving this unruly mess of further examples is one of the problems that yields short-sighted and incomplete analyses of the fallacy. Regardless of that issue, the lesson of this historical story, we think, is that the Watts model of logic text, followed by De Morgan, Whately, and Mill set the standard for a twentieth-century boom in informal logic textbooks. And so our history arrives at the present day. Still the straw man is struggling to break free. Beardsley (1950) is a kind of template for the genre. His is the first appearance, outside of Watts's oblique reference, of the term "straw man" in a textbook that our unscientific survey turned up. He mentions it explicitly in the discussion of oversimplification and distraction:

> In a dispute, the oversimplifier likes to set up a straw man and knock it down. He takes the weakest arguments for his opponent's view, states them in an absurd and extreme way, and then proceeds to dispose of the issue by refuting them easily.
>
> (143)

Now, of course, our speaker must use the straw man without drawing attention to the fact that it is an oversimplification. Else, we see a clear version of the effectiveness problem for straw man arguments. Interestingly, the straw man belongs, for Beardsley, among the appeals to emotion (which he doesn't really call fallacies). They are, in his telling, causes of imbalance or lack of neutrality in interpretation. This is an important development, since not only does he invoke a triadic scheme for the presentation of the straw man (speaker, target, and audience) but he takes it that the audience is primed to accept a negative and inaccurate depiction of the target. This, again, we think is a real insight—polyadic dialectical arrangements can be rife with what we'll call *the hermeneutics of antipathy,* and this is an important explainer for how straw man arguments are effective.

2.3 The Straw Man in Contemporary Textbooks

The straw man is a common fixture of informal logic and critical thinking textbooks nowadays. It's not universal, however, as some very popular texts omit it: Groarke and Tindale's *Good Reasoning Matters!* and Copi and Cohen (at least until 9th edition) are notable examples. By and large, the discussions of the straw man differ little from the standard treatment of fallacies one usually encounters in the texts. They span a page, or even less, but rarely more. They contain a brief description, an example, and often some recommendation for detecting or avoiding the straw man. Exceptions to this are Scriven's *Reasoning* (1976), Johnson and Blair's *Logical Self-Defense* (2006), and Govier's *A Practical Study of Argument* (1997), which contain longer treatments. For our purposes here, our survey of the history of the concept will finish with a look at how the straw man is presented in logic and critical thinking textbooks, and we will move forward with the contemporary theoretical accounts of straw man arguments in Chapter 7 once we have presented our view in the intervening chapters.

In these textbooks, there is broad agreement on the basic description of the straw man process. Typically, it's a two-step affair featuring first a mischaracterization and second a refutation of the mischaracterization. The mischaracterization is often described vaguely, but occasionally it's given some detail. Flage (2004) for instance, writes that in committing a straw man, "you either suggest that the argument presented was enthymematic and attack the allegedly suppressed premise or you misrepresent the conclusion and attack the alleged conclusion" (340). Baronett includes it among "fallacies of emphasis" (which includes composition, division, among others) and holds that it "often occurs when someone's written or spoken words are taken out of context" (148). The distorted version is then "easily refuted." Along these same lines, sometimes it's a matter of the imputation of standpoints by inference, as Arnauld and Nicole and De Morgan described earlier. Blair and Johnson, for example, discuss straw manning in the "extrapolation from a stated position." A person doesn't thereby *misrepresent* someone's explicitly stated view as they *attack conclusions thought to follow as a consequence* from that view. It seems true that this is a kind of misrepresentation, because someone is alleged to hold a view that they do not hold, but the mistake is actually a different one. The question, after all, is whether those commitments really do follow. When the imputed consequences do not follow, the mistake will be a *logical* rather than a representational one.[7] Interestingly, Moore and Parker (2004) include the imputation of fictitious and nefarious motives among the means of straw manning: "often a part of a straw man fallacy, is... presuming to read the minds of an entire group of people" (178). This is an insightful recommendation and it explains why straw man arguments so often have purchase. The imputation of a secret motive is permission, as it were, to replace the explicitly stated argument with another one. In this case, we see room for varieties of misrepresentation. We take this up at length in later chapters when we address hollow manning (complete fabrications) and the puzzle of effectiveness for straw men.

Whatever the means or method of misrepresentation—wrenching from context, misinterpretation, drawing of extreme conclusions—the straw man, following the tradition of Arnauld and Nicole, is regularly classed among the fallacies of relevance. To put it schematically, someone concocts, by whatever means, a misrepresentation of another's position and then criticizes it. The defeated replacement view is not the same as the original view, and so therefore it is irrelevant. They have made a case against an argument that is not on offer.

One challenge of explaining relevance comes across in the use of examples. Here is one from Hurley and Watson (2018):

> Mr. Goldberg has argued against prayer in public schools. Obviously Mr. Goldberg advocates atheism. But atheism is what they used to have in Russia. Atheism leads to the suppression of all religions and the replacement of God by an omnipotent state. Is that what we want for this country? I hardly think so. Clearly Mr. Goldberg's argument is nonsense.

This certainly recalls the view, discussed above, that fallacy examples just don't cut it. In order to fit the form, the sample text has to contain two arguments. The first one, in order to give the refutation a veneer of credibility, in this case stated somewhat generally (enthymematically as Moore and Parker have it). There is some argument to address. We're not actually told in this case what the argument is. It's hard to imagine being presented with the specific argument, for it would be laughably obvious that the person has misunderstood it. This also underscores a point we will discuss at greater length in Chapter 5. None of the texts we surveyed took seriously the idea that the straw man takes different forms (how the target is misrepresented and who is the audience for the misrepresentation are the two most important, by our lights). And the irony is that their representations and examples themselves represent a wide variety of means and forms. Crucially, the different forms suggest that the main dialectical issue with the straw man is not actually relevance but something else. This primarily is because the arguments criticized do bear on the issue, but are bad and criticized as bad. That *is relevant*, and so what gives? We need a more nuanced notion of relevance here that reflects the disappointment in what's misfired. This is why Christopher Tindale introduces the concept of "dialectical relevance" (2007: 24) to capture staying on-task in the critical conversation with one's interlocutors. The example from Hurley and Watson above also highlights another much overlooked problem, what we've called *the puzzle of effectiveness*. It is easy to see how the *ad hominem* is effective even in a made-up example, but it's hard to see how straw man arguments could be. At bottom, we think, straw man arguments are very ill-suited for the standard informal logic kind of approach. This is why, if we were to point to a successful account of it among the textbooks, we'd point to Govier (1997) or Johnson and Blair (2006). They provide robust context in their examples and they discuss the broader dialectical backdrop. Johnson and Blair, for example, provide a real live example (though we've argued that this isn't strictly necessary) along with a reading comprehension quiz on that example. This way, the reader is oriented to the accuracy of the position that will be straw manned. Then they provide the straw man version of the first argument. Another virtue of this approach was already signaled above by Whately: the straw man takes advantage of our sloppiness as an onlooking audience to the exchange. We read superficially, or listen inattentively; then we're easy marks for the misrepresentation. Again, as we think it, the issue is *who* the audience is for the straw man that will be part of the explanation for how it works as eliciting assent or silencing a party in the debate.

For better or for ill, but mostly for ill, our short excursion into the problem of fallacy names, pedagogy, and textbooks has revealed the sometimes very serious limitations of these documents and theoretical programs on the straw man phenomenon. One major limitation comes out: the straw man is included among the fallacies of relevance, because this is taken to be the salient *logical* failing to the argument. We, again, think this is an incomplete account of what goes wrong, because there can be virtuous misrepresentations since refuting a bad argument on a matter of dispute is a relevant contribution. More on that later. The point now is to note that this is the history of theorizing about the fallacy since Arnauld and Nicole and, in some very attenuated sense, Aristotle. It certainly is true that the substituting one argument for another and then criticizing the substitute would be an argumentative failing. But that's not a failing in a strictly logical sense. We need another dialectical notion. Further, it's a fallacy, so described, that would fool very few people—perhaps only the most inattentive or dishonest. But that we merely need to pay attention to what we hear and read is not a particularly gripping lesson in reasoning. The real failing, we think, is a dialectical one. For an alternative account of this, we can turn back to one highlight in our brief history of the straw man.

As we noted above, Whately's treatment of the straw man was particularly theoretically nuanced. We can see this now in comparison to the very brief tour of textbook treatments of it. Whately notices that oftentimes the targets of straw manning are downstream audiences, at one, or perhaps even two degrees removed from the original statement. He also noticed that the question of representation does not have to concern simple first-instance distortion—that some arguer A takes B out of context and then refutes B and B accepts the refutation. A may rather select the weakest parts of a case and do a really bang-up job of refuting them. This has the effect in the mind of some onlooker C that B's case isn't of the quality that it had initially seemed. Importantly, this doesn't have to entail the *defeat* of B. Rather, it might be enough merely for C to *withhold judgment* on the dispute between A and B. In this case, then, the straw man is not a matter of relevance or diversion at all. The weaknesses of B's case are relevant and A's arguments aren't misrepresentations. Nonetheless, A has warped the dialectical terrain in a way that needs to be explained in more detail, and it's here that the fallacy resides. We will have more to discuss with Whately later in Chapters 3 and 4.

Another shortcoming of the standard textbook account of the straw man is harder to describe. Since the straw man is a kind of souped-up *ignoratio elenchi*, it is one failing among many. In a sense, categorizing it this way, along with many other cases of incompetent argument (as imputed by *ignoratio elenchi*), does it a real injustice. Now we might be biased, as we have a book to publish on this topic, but there is something unique about straw manning in particular. Now, it is not wrong to say that it has the failings the textbooks correctly ascribe to it—relevance, representation, and so on. But what's unique about the straw man is that it is a kind of broader failure of rationality. This is not true, as we have mentioned, of other fallacies often placed alongside it. The *ad hominem*, historically and traditionally a fallacy of relevance, fails in a somewhat self-enclosed way—it takes something that isn't relevant and makes it seem so. Or take asserting the consequent. You're just misusing an intuitive rule bearing on the form of a conditional—get straight about sticking with "if" clauses, and you'll pretty much fix the problem. The problem of straw manning is much harder to avoid, and to see it as strictly about how we can err when we not only think about what we think about but we think about how others think about what we think about, and further how they can get things wrong when they think about how they and we think about things. It's an error that comes out of thinking about alternatives, and reasoning about those alternatives. To reason, after all, is to consider contrast cases, what the alternatives for a decision or a set of options for a view are. Our reasons sort those as good ones and bad ones, but if some of those options are not properly represented in ways that capture what makes them plausible or in ways that distort how they can handle a complication, this is where the straw man happens. More than the connection between contrastive reasons and how straw manning works from there later in Chapter 6. But our point is that the presence of contrast cases and how we think about them properly is psychologically demanding—this is part of the reason why disagreement generally and handling an ongoing argument is so hard—you have to keep a book, of sorts, of what's been established, what you and your interlocutor are on the hook for defending and explaining, and so on. And then we reason about not only what we've reasoned about, but we reason about the reasons we've logged in the book. In dialogue, because we must have a representation of what the alternatives are and how we've reasoned about them, the danger of straw manning is everywhere.

Lastly, as straw manning concerns the proper management of contrasting reasons, it is a very hard problem. The standard recommendation of textbooks to be careful, or to be accurate, or even charitable is certainly welcome, but insufficient. For, as we will see in subsequent chapters, charity can involve a dialectically and morally pernicious kind of misrepresentation. And accuracy can be easily abused, as we will show it is in the case of weak manning. And being careful about not accepting bad reasoning is exactly how the straw man argument works on us. Nobody ever thought that reasoning well was going to be as easy as following recipes from a textbook for sophomores, right?

2.4 Conclusion

We now bring our short history to a close. But as we close, let's return to the distant origins of the straw man. From its humble beginnings in Aristotle's *ignoratio elenchi*, the straw man is now a fixture of informal logic textbooks. It has sloughed off its initial form and now proudly bears its own name, or names (as we're going to argue). The contemporary picture is fairly clear. But what of the *ignoratio elenchi*? What we now know as the *ignoratio elenchi* from standard informal logic-style introductions to logic texts, such as Copi and Cohen, Hurley, or Baronett, is what is left after everything else is peeled away and given its own classification. It is a vaguely defined kind of dumping ground for problems of relevance or dialectical incompetence not otherwise specifically defined. Here, for example, is the latest edition (13th) of Hurley and Watson:

> This fallacy occurs when the premises of an argument support one particular conclusion, but then a different conclusion, often vaguely related to the correct conclusion, is drawn.
>
> (2018: 139)

They even admit that their definition is rather general:

> *Ignoratio elenchi* means "ignorance of the proof." The arguer is ignorant of the logical implications of his or her own premises and, as a result, draws a conclusion that misses the point entirely. The fallacy has a distinct structure all its own, but in some ways it serves as a catchall for arguments that are not clear instances of one or more of the other fallacies.
>
> (140)

One might suspect that the *ignoratio elenchi* is included in these sorts of treatments for a sense of completeness: for again, unhappy arguments may all be unhappy in their own unique way, and thus the need for having a name for the new ones or the ones that just don't fit in the other categories.[8] This is what remains. And so while we wait for a new christening, *ignoratio elenchi* will have to do. Unsatisfying as this might be, it's not that far off from the ancient origins of this fallacy. What it actually seems to be nowadays, once one has a look at even some of the history of it, is a kind of an appendix or a vestigial tail of fallacy theory.

3

Straw Men, Weak Men, and Hollow Men

The rough idea lurking in critical thinking and informal logic textbooks is that the straw man fallacy is a kind of misrepresentation of an opponent's argument as weaker than it actually is and so easier to criticize. The standard story (again, we'll cover the scholarly literature in Chapter 7) has it that this is a logical (rather than, say, a dialectical) problem. To be precise, it's a failure of relevance, because the misrepresentation of someone's case has no real bearing on its actual strength. To commit the straw man is, in other words, to infer badly. As fallacy of relevance, the straw man belongs in the same class as the *ad misericoridiam*, the *ad baculum*, and the *ad hominem* (among others). We're going to argue here that these two features of the standard picture—that the straw man's primary argumentative vice is constituted by misrepresentation and that its logical failure is one of relevance—don't provide an adequate account of the straw man's fallaciousness. There is much more to the fallacy of straw manning than simple misrepresentation. For in the first place, it is an oversimplification to say that misrepresentation on its own is always bad. Some misrepresentation, indeed some of what we might consider standard straw man misrepresentations, don't seem to be bad at all in a number of very common contexts. Further, as we'll also show later, there is more to straw manning than misrepresentation, or at least misrepresentation in the first instance. For one can straw man, we'll argue, by *being scrupulously accurate* about a particular opponent's thesis or by focusing in on one of their arguments. This is what we call the *weak man*. But it's also possible to straw man without distorting (and so misrepresenting) an argument at all, as is the case in the *hollow man*. One can just make things up about another's views, and instead of distorting anything they've said, one can just disregard it. Another central contention of ours, which we'll also begin to develop here, is that the straw man isn't best understood as a problem of relevance, like the *ad hominem*. Standardly, the *ad hominem* is fallacious because someone's character isn't relevant to whether their argument is valid or not. The straw man, however, is not a problem of relevance because it's not really a logical problem at all; it's a dialectical problem, one rooted in the give-and-take between arguers.

Now to the plan of this chapter (and to some extent, the next one). This chapter's main focus is to describe three varieties of straw man. We'll start with what we call the *representational form*, or standard straw man, because this is by and large what most people are familiar with. We will then identify further forms in relation to this, as it will become clear that what the representational form means to capture actually admits of a startling amount of complexity. The second form, unlike the one just described, does not involve distortion in the first instance, but rather an uncharitable selection of an opponent's weakest arguments for criticism. Call this the *selection form* of straw man, or the *weak man*. Further, we will consider yet another very common but little discussed variety of straw man that we will call the *hollow man*, instances where the critic makes up an entirely fictitious argumentative opponent. There's a sense that it's hard to call it a *misrepresentation*, since it's not clear what or who it is a representation of. Once we get a handle on these basic forms, we'll evaluate some cases, discuss puzzles, and finally begin discussion (which we'll continue in later chapters) of their fallaciousness. We will close with a discussion of whether there can be non-fallacious cases of any of these species of straw man arguments. And the important takeaway, then, is that if there are non-fallacious versions even of misrepresentations in argumentative exchange, then it seems that it can't be the misrepresentation that explains what makes the straw man fallacious.

3.1 The Straw Man and Weak Man

You'll pardon us if we get a little schematic—there's a lot to track and this makes it easier. The most basic straw man scenario involves, at minimum, a target B with a position p with arguments x, y, z (more or less) and an arguer A who acts as critic of B. A is the perpetrator of the fallacy here; B is the victim or target. Generally, in a straw man instances, A misrepresents something about B's case. A quick inspection shows that this can happen in two different general ways. In a first variation, A listens to B's arguments x, y, and z for p and replies by describing p* rather than p. A then follows with a devastating critique of p* because p* is much less defensible than p. Much to B's chagrin, p* is not B's position at all, so A's criticism, however incisive, of p* is irrelevant to the cogency of B's position p. Simply put, in the first variation A distorts B's conclusion. In a second variation of the basic representational form, A alters some aspect of B's arguments x, y, z or their logical connection to p. In this scenario, A misrepresents the quality of B's inferences to p.

So much for a very bare bones description of the representational straw man. It turns out that A's options for misrepresenting B are quite vast, a fact which is crucial to understanding the straw man as a fallacy. On the first variation, A can restate B's conclusion such that it's much weaker than it seems to be. This might involve, as Arnauld and Nicole suggested, imputing a commitment to B that is "very far from his meaning" or by drawing out consequences thought to the A to follow from B's view, though not expressly argued for by B (1850: 242). Alternatively, A might state B's conclusion in a more certain or uncompromising form—where B was tentative, A might represent B as decisive, thereby altering the strength of the inference from x, y, z to p. The options here are pretty extensive, because, again, unhappy families are all different in their own way. Lots of things will be different from the unfortunate B's original thesis. A might even repeat the actual thesis, in an oafish voice or, what is perhaps even worse, *an evil sounding voice*, as in the *modus tonens* (Aikin and Talisse 2008). If the argument is in print, perhaps **A might even quote B's argument in Comic Sans**. There are so many good options.

On the second variation, A might misrepresent the *quality* of the reasons—the arguments x, y, and z—for the *correctly* stated conclusion. Or the dastardly A might misrepresent the *quantity* of reasons. As Flage (2004, see Chapter 2) noted, A might present B's reasons as enthymematic and so "suppressing a premise." Clearly, A's options for distorting B's reasons are, as just mentioned, as vast as the catalog of things that can go wrong, or right, in an argument. Consider (again) that A might simply weaken a premise by stating it more strongly than the usually careful B would normally have stated it; simply change a quantifier from *most* to *all* and, voilà, B's argument has been undermined. Alternatively, A might also omit some of B's many arguments x, y, and z for p, thereby making B's reasons appear inadequate. Our interest here is not with the various means of distortion—again, those are as limitless as the grains of sand on the beach. Our focus here and in the subsequent chapter is with the broader dialectical context of straw manning.

If you've ever been in B's shoes, and we can assure we both have, you know how annoying this is. But annoyance isn't quite fallacy theory, so next we ought consider what makes A's arguing fallacious. Once we do this, we'll be in a better position to see how A's options for straw manning include more than first instance distortions of B's arguments—and, as a consequence, why straw manning is an interesting puzzle. We'll also be able to appreciate the conditions that make A's straw manning effective and B's options for correction so limited. In later chapters, these features will show that the straw man belongs to a peculiar class of underdescribed or perhaps misidentified fallacies—fallacies of meta-argumentation.

As we've discussed in Chapter 2, the straw man is usually considered a fallacy of relevance. There are two significant historical reasons for this. The first is that the straw man came from the *ignoratio elenchi*, which was a general stand-in concept for a relevance problem. Also, even though the fallacy theoretic from which the straw man emerges had its origins in Aristotle's account of a question-answer dialectical competition, subsequent theorizing often (with some important exceptions) focused on characterizing the *logical* or *inferential* failure in question. Viewed from the more narrow lens of informal logic, it makes intuitive sense that the problem with straw manning is one of irrelevance. Speaking broadly, A's version of B's conclusion p or arguments x, y, z for p isn't really B's conclusion or B's argument, so A's refutation is irrelevant. So, in assembling or constructing a straw man, and then knocking it down, A infers badly. Or perhaps B infers badly if they take A to have successfully refuted them. Either way, someone is inferring badly. Unfortunately, this general account leaves out all of the interesting stuff. It's only a tad better than calling all fallacies *non sequiturs*, because it's always true that the conclusion doesn't follow. It also suffers from what we call the effectiveness problem, because it's hard to see how such obvious misrepresentations could bamboozle anyone, even the hapless B.

A standard way of explaining informal fallaciousness is to assimilate an argument to another well-described fallacy. This is a handy method in that it's ontologically parsimonious and epistemically cheap. Even though *Anna Karenina* tells us that unhappy families are individuals, it is still true that you can generalize over cases where spouses fall for someone named Vronsky, or those where the paterfamilias is struck with a persistent debilitating illness. In this spirit, one option, noted by van Eemeren and Grootendorst (1987: 286), is to consider the straw man a version of the *ad hominem*. There is a lot to recommend this. Consider again the two variations of representational straw man. In the first A alters B's conclusion p to make it less defensible, perhaps by making it more extreme or by replacing with something else entirely.

By doing this, A suggests that either that B is *incompetent* for not realizing the obvious logical implications of their own view or that B is *dishonest* because they don't offer their *real* conclusion. A similar analysis applies to altering B's arguments. In a way, this is a more subtle version of the name-calling personal attack—the thinking man's *ad hominem*, as it were. For, stressing the wrongness of B's conclusion, or B's failure to grasp what their conclusion actually is (or entails), is merely a means to show that B isn't worth engaging with. The representational straw man is a kind of roundabout way of calling B dumb. The virtue, by which we mean the vice, of doing it this way is that gives the semblance of justification, as B did—so it seems at least to the incautious—make a silly claim, didn't they? In arguments, after all, we're at the mercy of the evidence. This is certainly true, but it's trivially true of all accusations of argument failure. This is something we warn our students: when your argument comes in for criticism, it's going to follow that you've failed to notice the grounds for that criticism. There are other candidates for assimilation. Not to belabor the point here but some theorists have wondered whether the straw man fallacy so described is a sub-species of hasty generalization (Chase 1956: 40), or a failure of internal proof or a *secundum quid* fallacy (Vernon and Nissen 1968: 160; Walton 1992: 75–80). There are potentially other options suggested by the various classificatory schemes found in the many accounts of the fallacy in the textbooks, but this isn't our real interest here.

Aside from assimilation to other well-theorized fallacies, there are some other ways to characterize the fallaciousness of the straw man. The straw man is a fallacy because it marks a pragmatic failure of argumentation, a failure to connect and coordinate with another person in the exchange. A's straw man argument against B undermines the goals of critical discussion because the resolution of such critical exchange requires that parties argue responsively to one another. That is, A's setting up a straw man of B's view is a failure to actually engage with B.[1] Consequently, what makes the straw man a fallacy, so far, is that a speaker who erects a straw man advances an argument that misrepresents to their advantage what is taken to be the current dialectical situation. The crucial element of this misrepresentation that distinguishes the straw man from other misrepresentations is that the strength of the opponent's case is not reflected by the arguments the speaker attributes to the opposition.[2] To put this another way, the straw man makes the entire critical discussion a waste of time since the participants are not engaging with each other. While this approach has some virtues, like its recognition of the fundamental dialogical nature of the straw man, it misses out on what makes the straw man an effective strategy and why, puzzlingly, people continue to employ it. We'll have more to say about this in Chapters 5, 6, and 7. For now we'd like to move onto an important variation of the basic form.

If opportunistically misrepresenting the current dialectical situation by responding to weaker arguments than those given is the initial vice of the straw man, the fallacy admits of forms other than the one presupposed in the standard analysis. Consider that the ever-devious A can misrepresent the current dialectical situation not with regard to what their current opponent B says, but with regard to the variety and strength of opposition to their view, of which B may only be a questionably able representative. For example, let A, in arguing for their position, survey objections to their view. They take up their opponent B, *correctly* recount B's objection, and then *legitimately* refute it. So far, so good for A. But crucially A has ignored C, whose objection to A's position is like B's, only stronger in that it avoids the extravagance or imprecision of B's objection. Now comes the trick: A then concludes that they have successfully defended their view against all comers.

In such a case, A clearly distorts the dialectical situation in responding to weaker arguments than given by their opposition taken as a whole. In taking up with only B's objections, A pragmatically implicates that A is taking up with the best ones (again, on the presumption of proper resolution of the critical exchange). And it is here that it is worth pausing to note the meta-argumentative background to the straw man—there is an argument about the overall state of dialectical play given in the response to a range of bad arguments. We hold this is an important feature of straw man cases—that of reasoning to *close an argument*. To be clear, A does not misrepresent B's position or argument in the exchange, A nevertheless straw mans their *opposition* more generally by refuting only their weakest opponents. A has erected a straw man, but they have not *misrepresented* the objection they seek to refute; instead, they have erected a straw man by *selecting* a relatively weak version of, or inept spokesperson for, the opposition to their view. Hence there are two forms of the straw man: the *representation* form, which is the one usually captured in the standard analysis of the fallacy, and the *selection* form. We propose, to make it clear that we see the metaphor here differently from an *opponent of straw*, what our arguer has done is *select the weakest* of the opposition as her B. So we propose the updated metaphorical name, the *weak* man.

On this description, the *selection* form of the straw man fallacy bears a strong resemblance to the fallacy of hasty generalization. Standardly, a hasty generalization consists in drawing a general conclusion from an inadequate sample. Say you plop down in a foreign country for the first time, where you see someone wearing a pumpkin, and you thereupon conclude that pumpkins are everyday fashion. Well, hate to break to you, but you've hastily generalized from Halloween. In this case, consider that A cherry picks B as an opponent on the basis of B's dialectical incompetence, implausible commitments, or flimsy arguments. A then refutes B and implicates that refutation of B is a repudiation all similar dialectical resistance. However, the similarity with hasty generalization only goes so deep. Avoiding the selection form of the straw man does not require all arguers to overtly respond to every potential challenge. Surely we must recognize that some challenges are too trifling to bother with; after all, life is short and crackpots are a plenty.[3] The problem with this case of A's straw manning by selection is not that not every objection fits the mold given in A's argument (again, A doesn't have to address every argument) but that A does not address the better arguments in the opposition. That is, the problem is not a *quantitative* problem with the argument (as the hasty generalization analysis would run) but a *qualitative* one. Although A is not obliged to address every form of objection, they are obliged to address the best they can find. In addressing B's arguments only, A implicates that B is the best or most relevant voice of their opposition, because one resolves disagreements most effectively when one attends to and addresses one's opponents' best arguments. If A represents their view as defended on the basis of a refutation of B, then it must be reasonable to conclude that the opposition's best and most influential voices do no better than B.[4] This, again, is the meta-argumentative element of straw men arguments, focused on closing the issue with these discussants.

Like standard representational straw manning, weak manning affords the devious arguer, such as our friend A is, with a long menu of options. To start, A could misrepresent the overall form of B's argument but respond only to selected parts of it. For example, B's argument for p may be a convergent argument, one with a number of individual arguments that aggregate to a larger argument, stronger than any of the component cases. Alone, the individual arguments may not carry the day, but together they may make a substantial case. So B's *full* argument for p is argument a (x and y and z), and the proper way to understand B's case is, again, as x, and y, and z yielding the cumulative case, a. Now comes the ever-critical A who may latch on to argument x, quote it and interpret it exactly correctly on it individual level, and then argue (or just implicate) that B's case is weak. A could then turn to y, and then to z serially and individually. In each case, A can note how weak the argument is on its own. At the end, A will then note that B's argument, on closer inspection, doesn't hold up. The problem, of course, is that B's case is cumulative and so doesn't stand on any one of the individual arguments but is a convergent case. For example, in philosophy of religion, the case for atheism can be considered as a cumulative or convergent case. The argument from evil, the failure of theological arguments, the incoherence of the divine attributes (e.g., paradoxes of omnipotence), and the power of debunking arguments makes the overall case against the god of traditional Western theism. Individually, the arguments themselves are not particularly probative—for example, the argument from evil has the out for skeptical theism and so does not offer a full case for atheism outright. And the failure of theological arguments does not by itself show that atheism is true, as such an inference would be a form of committing the argument from ignorance. But overall, once shoulder-to-shoulder, the convergent case against theism can be very strong. Rebutting each argument individually as insufficient for a case against theism is a mischaracterization of what the dialectical situation is for theism and atheism.

To return to the general options. It is clear that opportunities for the selection form of the straw man are legion. All of them require that there is more than one argument in play. There may also be more than one arguer. A might claim that their refutation r defeats *all* of B's arguments (x, y, z) for position p, because the argument A selected was sufficiently representative of B's arguments for p. As we have pointed out, this resembles the hasty generalization, since it picks an unrepresentative sample to stand for all. Or, alternatively, the weak man form of argument may be less a form of quantitative error as seen as a problem of generalizing induction, and more a qualitative error. That is, in selecting B's argument x and attending only to it, A has, again, implicated that x is the best or most powerful argument that B makes. That is, if A is taking the time to respond to one of B's arguments, then A must take argument x as providing some comparative measure of rational resistance to A's preferred position. If A passes the other arguments over with silence, A implicates that they are not worth responding to (or at least do not have the urgency that x has). As a consequence, the implication of responding to an argument is that one takes the argument as one *worth responding to*, namely that it is good, or at least the best the opposition can do. Consequently, in performing the selectional form of the straw man, A has inverted the dialectical terrain. A takes B's worst argument, x, and in responding to it only, implicates that it is the best that B gives. And so, given that A can respond to x successfully and does not answer B's other arguments, the implication, again, is that A may not have responded to all of B's arguments, but defeated the *best*, which is, for limited purposes, good enough. And so, we see the problem here is not with misrepresenting the B, but with bad reasoning about the reasons under consideration. The straw man is a meta-argumentative fallacy.

Importantly, the selectional form of the straw man also has resonances with *ad hominem* abusive or a kind of negative ethotic argument. The weak argument's selection, its consequent easy refutation, and A's implication that this is the best argument that B gives, undermines our confidence in B's reasoning power, so that their authority as an arguer is called into question in virtue of the one egregious, or at least poorly representative, example. Notice that this form does not involve the direct or first-instance misrepresentation of an opponent's argument as it does the more global failure to exercise charity in selecting which of an opponent's arguments to address. But nonetheless, the selection form does engender a misrepresentation in the "second instance" for it is meant to suggest that either B's overall argument is weaker than it appears or that B is a less worthy interlocutor (the consequences of this for the effectiveness puzzle will be explored in Chapter 5).

As we can see, the paradigm case of the weak man involves the "first instance" selection among arguments B actually makes for position x. But the weak man might occur in a larger context as well. For example, in responding to objections to his view, A might select the weakest arguments from among arguments of several *different* arguers (B, C, and D). In this case it's not the weaker or weakest of one interlocutor's, B's, arguments but the weaker or weakest of arguments against A's views which is a position which B holds (and which C and D hold as well). Schematically, the distortion may be presented as follows. B, C, and D all hold that p, but they hold that p on the basis of a wide variety of arguments. B, perhaps, is sophisticated, and they hold that p on the basis of arguments x, y, and z, which, by the standards set by the state of the dialogue, are good arguments. C and D, however, are not quite up to snuff, and though they get B's arguments, when they try to give them, they muck them up. C holds that p on the basis of distorted and more criticizable arguments x^*, y^*, and z^*. And D just holds p on the basis of x^*. A does not need to distort standing arguments for p, now, as those who argue for p have done that work for them—all they need to do is find and pick on the members of the opposition that are more mistake-prone or less careful.[5] A may, in fact, be very careful to get C and D's argument x^* *just right*, so as to *highlight* just how bad the case for p is. And, again, this is where the distortion happens—though x^* is the most widely given argument for p (C and D both give it), it is not the best case for p. But A's audience, unless they were antecedently aware of B's case, would not only never know that, they would positively believe that *the case for p generally* is as weak or weaker than x^*.

The selection form of the straw man fallacy is vicious because it is posited on a misrepresentation of the variety and relative quality of one's opposition and then an inference about the plausibility of a useful critical discussion continuing. One question is how such arguments succeed in eliciting assent. This, we've termed *the effectiveness puzzle*. Generally, straw man arguments depend not only on A's misrepresentation of her opponent's commitments and arguments but on their audience's inexperience or ignorance. In the case of the representation form of the straw man fallacy, A's argument against B depends on A's deliberate presentation of p* (or a*) as equivalent to B's p (or a). B can correct this misrepresentation only if they keep track of their own position and arguments for it, and if they can recognize and articulate the difference between them and A's corrupt versions. If B cannot, then A succeeds, but only in a sense (more on that in Chapters 5 and 7). Additionally, if A's audience cannot keep track of B's commitments (or has no interest in doing so), then A's representational straw man succeeds. By contrast, in the selection form of the straw man argument, A *correctly* presents B's argument and *legitimately* refutes it, but they fail to countenance stronger objections from other sources.[6] In so doing, A implicitly presents themselves to their audience as having successfully defended their view against the best cases and as having thereby *established* their view. Unless their audience is familiar with the better counterarguments proposed by A's opposition, then A succeeds in winning their assent.

The two forms of straw manning, then, take advantage of two different failings in their respective audiences. The representation form of the straw man argument depends on the audience not detecting the difference between B's arguments (x, y, z) or position (p) and A's misrepresentation of them with x*, y*, z*, or p*. The audience must be inattentive in the sense that any nuance of p or sophistication of x, y, z does not register or is deemed inconsequential.[7] In the selection form of the straw man, however, A's argument depends on the audience being unaware of the variety and relative quality of opposition to A's position. The audience, by contrast, can be very attentive to the details of B's position and have the listening virtues not present in representation forms of the straw man. However, the audience is not knowledgeable about A's opposition, so A sets the terms for argumentative success by default.

Correlatively, the requirements for correction in the case of the selection straw man will be different from those of the representation straw man. First, showing that A has ignored some other versions of the opposition is necessary. This may take the form of either showing that though B gives argument x*, others give better versions of x that do not have B's problems, or one may show that though B's argument x* is relevant, there are entirely different and more pressing arguments from other sources. Second, an objector must demonstrate that these ignored arguments are superior to the one to which A attends.[8] The general difficulty is that what must be demonstrated is that A has misrepresented the overall dialectical situation and most importantly the quality of her opposition (viz., the state of the art in a debate), but there may yet be no consensus on what that dialectical situation is. However, even if it is an open question as to what the facts of the matter are as to whether B, C, or D has the best argument, A has unjustifiably simplified the situation by taking on B only. The consequence, then, is that correcting the selection form of the straw man fallacy requires more than an analysis of what some speaker says, but an education in the larger discourse A purports to be addressing. That is, in cases of the selection form of the straw man fallacy, A *relies on* the ignorance of her audience; if A is to elicit their assent with the selection form of the straw man fallacy, they must not be familiar with the best arguments made by A's opposition. In this way, the selection form of the straw man draws its success from the ignorance of its intended audience. The only correction is education of the sort advocated most famously by John Stuart Mill in the second chapter of his *On Liberty*. There, Mill argues that our understanding of our own position is directly proportionate to our understanding of those of our opponents.[9]

3.2 The Straw and Weak Man in Practice

3.2.1 Chicken Little Liberals

Once upon a time, Chicken Little freaked out when hit on the head with an acorn, and called out, "*The Sky is Falling! The Sky is Falling!*" Everyone goes berserk, but then they see it's just an acorn. Chicken Little then retires and admits to having overreacted, and things return to normal. The end. Calling someone a "Chicken Little," then, works as a form of analogy. One sees someone reacting strongly to something, perhaps that it forebodes something worse, and one then points out that they are overreacting because they don't see the situation clearly.

It was a common feature of contemporary American political culture during the Trump Presidency for people to think and say that Donald Trump is a danger not just to America's prosperity and safety but to the world's. He's an authoritarian, he seems to have (or at least there's the accusation that he had) colluded with another state to secure his election, and he seemed to be a general nincompoop who surrounds himself with avaricious doofuses without a care for democratic norms, environmental protection, sustainable governmental spending, and international goodwill. That makes him dangerous as the president of the United States. Heather Wilhelm at *National Review Online* had had it with the doom-saying Chicken Littles with regard to the Trump presidency.[10]

> The unprecedented volume of apocalyptic media pronouncements that Trump has inspired is unhealthy …. How many times can one presidential administration end life as we know it?

The coverage of the Trump administration is "crazed and breathless" and bent on spurring your outrage or stoking your fears with predictions of doom. Chicken Little apocalyptic journalism from what they see as the Trump-hating Left. But Wilhelm has a counter to this:

> [C]ongratulations! If you're reading this, it means you're still alive, and have survived the approximately 5,000 world-ending decisions that the Trump administration has supposedly made thus far this year. The Russians, at least as far as I know, have not yet taken over. Faced with budget challenges and various logistical challenges, including the fact more than 1,000 miles of our border with Mexico is actually a river, it seems that Trump's much-decried Great Wall of America could be slowly shuffled off into the "it seemed like a good idea at the time, but maybe not really" pile. When it comes to health care, congressional Republicans seem to be in the political equivalent of that one unlucky bumper car that gets stuck in the corner, no matter which way you steer. As Francis Fukuyama addressed the panic in Politico this week: "Trump's a dictator? He can't even repeal Obamacare."

The last line is funny—we'll give Fukuyama and Wilhelm that. But how is this a reply to the worries people actually had about the Trump administration? In fact, who had said that the readership of *National Review Online* would not survive two years of the Trump administration? The answer was: nobody. The regular complaint was that a Trump presidency would be, as Andrew Sullivan put it, "an extinction-level event for American democracy," but that was not the claim that people would not survive the Trump presidency. In fact, this is precisely what the force of the assessments of Trump that Wilhelm quotes say.

> "President Trump Risks the Planet," *The New York Times* declared on Tuesday, in the wake of Trump's executive order rolling back Obama-era climate regulations… With his characteristic flair for the dramatic, filmmaker Michael Moore took the idea to its logical endpoint on Twitter: "Historians in the near future will mark today, March 28, 2017, as the day the extinction of human life on earth began, thanks 2 Donald Trump."

Crucially, none of these As of Trump's policies were predicting the complete demise of civilization by the time of the 2017 publication of Wilhelm's article in *National Review*. As Lewiński and Oswald (2013) have noted about straw man interpretation, there is a standard of interpretation that has, for sure, penumbral boundaries, but there is a reasonable limit to how uncharitably to take a statement along these lines. What's crucial is that the Moore line quoted invokes *future historians*, even if it also invokes *the extinction of human life*. Yes, the rhetoric is overheated, but even with such apocalyptic invocations, it seems that someone still is looking back. And so, it is clear here we have a representational straw man argument, and it is worth noting that the force of it is not simply to answer misrepresentations of the Bs' arguments but to paint *them* in a particular light so that we may reason not just about their reasons, but them as reasoners, namely that they are not people with whom to have continued genuine exchange.

3.2.2 Kick 'em

In October of 2018, before the mid-term elections in the United States, Eric Holder, the former US attorney general under President Barack Obama, put a new spin on the familiar Michelle Obama quip, "When they go low, we go high." Holder's was that "When they go low, we kick them."[11] In the video of the reference reported at the *Washington Post*, Holder, after the quip, clarifies what he means by "kick" them:

> When I say we, you know, "We kick 'em," I don't mean we do anything inappropriate. We don't do anything illegal ... But we got to be tough, and we have to fight for the very things that [civil rights leaders] John Lewis, Martin Luther King, Whitney Young—you know, all those folks gave to us.

He means give it back with confrontational rhetoric, not actual violence. And Holder, again, is keen to clarify that point immediately after the quip. However, in reporting this quip on Fox News, the voice for the American Right and Republican Party, Holder's qualification is not reported. Fox News Commentator Sean Hannity focused entirely on the part about the kicking. The Holder quip is cut just before the clarification.[12] He frames and restates Holder's quip:

> Just look at the number of democratic leaders encouraging mob violence against their political opponents... When they go low, ditch civility. Kick Republicans, when they are on the ground, kick 'em.

He just plays the quip, not the clarification, note. And that's the key. Holder's expressing the view that political argument is high-stakes and hard-charging, so he's willing to sacrifice the high road during particularly important election cycles, precisely because he thinks it's clear his opponents have done so. That's what the metaphor of "go low" means, not "when they are on the ground," as Hannity has taken it, which means that they are injured. So the metaphor of the "kick" is the response to their "going low"—invoking the way the political battle would go. But it's all metaphorical about the rhetorical exchange. Imagine someone saying, after hearing another describing a coming debate as a "bare knuckle boxing match," they are worried for their physical safety. For sure, this would be some willful ignorance of how metaphorical language works. Political commentator Trevor Noah's *Daily Show* review of the selective quotation also revealed the additional irony: "Can we just acknowledge that by saying they're gonna get kicked, Sean Hannity and his friends are accepting that they're going low?"[13]

3.2.3 Straw Mika

A long-standing way to think of straw man argumentation is as misinterpretation or misrepresentation of what people said or what their arguments were. That's a version of the representational straw man. The weak man is a case of finding a member of the opposition that has a badly stated version of the view or a poorly constructed version of their argument and go after that.

There's nothing wrong with criticizing a bad argument, but what gets communicated with it is that you, in investing time and energy in replying to that bad argument, are not spending time on the better ones. That would be bad use of your time, so if you're doing the work of criticizing the bad arguments, they must be as good as they get. Again, that's the meta-argumentative core of straw-manning generally—in attacking bad arguments, you implicate that there aren't better ones to go after. Another weak man instance is that you take imperfectly phrased versions of an opponent's position and interpret them mercilessly. When we're speaking off the cuff, extemporaneously, we may not say everything just right. And so we, except when in full-attack mode, give each other some slack. That's a difference between spoken and written communication. And to interpret your interlocutor in the worst lights when they are speaking informally (and so, imprecisely) is a kind of selectional straw man.

Mika Brzezinski, co-host of the MSNBC show *Morning Joe*, said that the media's "job" is to "actually control exactly what people think."[14] Now, the context is that Brzezinski's line was a contrastive—that President Trump is trying to control what people think by pushing out the media. By way of his Twitter account, Trump is purportedly "speaking directly to the people," and so bypasses the "mainstream media" and its purported bias, but Trump's Twitter feed is full of misinformation and is not sourced. Here is the quote in full:

> I think that the dangerous, you know, edges here are that he's is trying to undermine the media, trying to make up his own facts. And it could be that, while unemployment and the economy worsens, he could have undermined

the messaging so much that he can actually control exactly what people think. And that is our job.

The key is to interpret what the "that" of "that is our job" is referring to, as the anaphor seems to be "controlling what people think," given the strict interpretation. However, given the complaint about Trump's Twitter feed standing in the way of responsible communication about how the government is doing its job, the more reasonable interpretation is that Brzezinski is talking about doing the reporting. Regardless, it's not a well-constructed series of thoughts, but that's how spoken communication goes, isn't it? The anaphorical relation between a pronoun or an index gets lost when we are talking, but a minimally charitable interpretation can correct for that.

In response to Brzezinski's line, conservative media reacted particularly negatively. Columnist Tyler Durden at *Zero Hedge* says Brzezinski "let slip the awesome unspoken truth" about what the media thinks they should be doing, which is controlling how and what people think.[15] And conservative outlet *Breitbart* made it a front page story for a day, with the implication that the imperfect wording is really a Freudian slip, letting everyone know that the media's objective is mind control.[16] Shortly afterward, Brzezinski went to Twitter to clarify the situation.

> Today I said it's the media's job to keep President Trump from making up his own facts, NOT that it's our job to control what people think.[17]

It's pretty clear that when folks have what we are calling a hermeneutics of antipathy already cued, they'll take something like this as evidence of letting a mask slip instead of a poorly phrased bit of intellectual pushback. So this makes it an interesting case of a mix between selectional and representational straw man—it's selectional, since they go after what she's actually said, but it's representational, since we need an interpretive attitude to take this as seriously a representation of her sincere position. So, in a way, a lesson about straw manning. If your picture of the opposition, after interpretation, fits the worst kind of picture you may have of them, you may be a straw manner.

3.3 A Puzzle about the Weak Man

To be sure, we have here offered a preliminary characterization of the selection form of the straw man, or the weak man. Much more analytical work needs to be done on this form of the fallacy, and more is in the offing to follow. However, our present aim in drawing attention to the selection form is not simply that of introducing a new dimension to a common, familiar, and thoroughly theorized fallacy. It is our view that the selection form of the straw man is a prevalent form of fallacious argumentation at work in contemporary popular political discourse. In fact, we hold that the prevalence of this form of fallacy helps to explain the curious confluence of two seemingly inconsistent phenomena in contemporary popular politics: (1) high levels of public ignorance about fundamental political matters, and (2) heightened attention to sources of political analysis and commentary.

The social scientific literature documenting public political ignorance is vast and will not be surveyed here.[18] Suffice it to say that the following estimation by Bruce Ackerman and James Fishkin is widely shared:

> If six decades of modern political public opinion research establish anything, it is that the general public's political ignorance is appalling by any standard.
> (Ackerman and Fishkin 2004: 5)

That Ackerman and Fishkin are the two advocates for a highly ambitious project of participatory democracy that they call "Deliberation Day" speaks to the force of the public ignorance findings; presumably, participatory democrats would attempt to downplay these findings, if there were an intellectually responsible way to do so. The evidence for the heightened attention to political analysis and commentary consists simply in the fact that the popular political book publishing business is now a billion-dollar industry. Even a cursory look through the politics section of a local bookstore will confirm the utter proliferation of treatises offering what profess to be detailed political commentary on the politics of the day, almost in real time. And this is to say nothing about the number of television programs, radio shows, and internet news sites devoted explicitly to current political affairs.

And so, we have a strange puzzle—though the prevalence of the representational and selectional versions of the straw man accord with a wide kind of ignorance of the debates, there is a large industry that is booming on publicizing these debates and a large audience for its consumption. It seems inconsistent on its face. One would expect that greater attention to political analysts and commentators—even highly partisan analysts and commentators—would result in a *decrease* of political ignorance. But the trend does not work this way. In fact, a recent study has found that increased attention to the media forms that tend to feature more by way of real time argumentation—namely, television and radio, as opposed to print sources—is *positively correlated with political ignorance*.[19] But this positive correlation between exposure to sources of purported political analysis and political ignorance is precisely what should be expected from a mode of public discourse in which the selection form of the straw man fallacy is prevalent. For it is the essence of this fallacy to cast the *entirety* of one's opposition in the terms adopted by one's *weakest* opponent. When the selectional straw man prevails, one's audience is convinced that *there is no intelligent opposition* to one's view, and thus *no forthcoming rejoinder from the opposition that could be worth attending to*. That is, though the straw man fallacy itself is not a form of hasty generalization, it does yield an analogous inference in its audience—those who hear straw man arguments take themselves to have inductive evidence for the stupidity of their opponents. Only a narrow and distorted view of contemporary political disputes can result.

Evidence that popular political commentary is governed in large measure by the selectional form of the straw man fallacy is garnered, again, by an even cursory survey of the popular political literature that can be found in any bookstore. In fact, one needs only to look at the *titles* of the bestselling books to get a sense of the extent to which the fallacy prevails. To cite only a few examples, conservative commentators claim that liberals suffer from a "mental disorder" (Savage 2005) and should be spoken with only "if you must" (Coulter 2004), while liberal commentators cast their opponents as "lying liars" (Franken 2003) who trade in "idiocy" (Black 2004). On all sides, the argumentative strategy is the same: The audience is expected to rely upon the author to present the opponent's view, the author presents what is in fact a more-or-less accurate depiction of what some of the weakest opponents have said, the author easily refutes the opponents, and then explicitly takes himself or herself to have shown that *all* extant articulations of the opposing view are as easily dismantled.

As we noted above, the selectional weak man fallacy depends upon the ignorance of one's audience; in order to succeed, one's audience must not have first-hand knowledge of how strong opponents respond to one's position. However, we now see that the selection form of the straw man fallacy, the weak man, serves to *perpetuate* if not positively *encourage* such ignorance. When it succeeds, it convinces one's audience not only of the correctness of one's view, but also of the absence of reasoned and intelligent opposition to it. The result is a popular public discourse of heightened passion and outrage that grows increasingly ignorant of what is actually in dispute. Under such conditions, a premium is placed on holding one's ground without regard to the reasons and arguments of those who disagree; that is, the result is a total undermining of argumentation.

On any view about the ultimate purposes and nature of public political discourse in a democratic society, the prevalence of a fallacy that undermines argumentation and encourages irrational tenacity must be seen as a threat to a properly functioning system of self-government. We have claimed that this form prevails in popular political discourse, with deleterious effects.

3.4 The Hollow Man

If we take the straw man fallacy to be centrally one about misrepresentation of the opposition's views and overall case and then making pessimistic inferences about prospects for further argument, then we might also envision the misrepresentation along a spectrum of accuracy of representation. The accuracy, we should caution, is "first instance" accuracy—whether or not someone's view is accurately or fairly represented. On this view, the most honest, as it were, form of straw man is the selectional form, or the weak man. Someone *does*, after all, hold the weak-manned position—in the first case, B actually does give the weaker argument, x, and in the second case, C and D do give argument x*. The dishonesty is the meta-argumentative emphasis put on this fact. Moving down the spectrum, next is the representational or standard straw man. The view is a caricature of an actual view, but it at least can be traced to an actual view. It may be a distortion of the dialectical situation, but at least it is addressed to someone who at least can clarify the situation. Moving down the spectrum we see room for another variation on the straw man, one where the opposing view is not a caricature but rather *a complete fabrication*. It represents no particular discussant, and it bears no relation to any view expressed. It is an unoccupied viewpoint. We might call this "the hollow man." More specifically, the hollow man consists in fabricating an imaginary opponent with an imaginary and impossibly weak argument, and then defeating the argument.[20]

Unlike the straw man, where the opponent is real but the argument is distorted, or the weak man, where both the opponent and the argument are real but the overall state of play in the dialogue is distorted, with the hollow man, the argument and sometimes the opponent is completely fake. We say "sometimes" because the hollow man, like the earlier forms of straw man, admits of two distinct varieties. One variation of hollow man consists in fabricating both an opponent and an opponent's view, merely in order to defeat them. We might call this the *extreme variety* of the hollow man. Oftentimes this is signaled by a vague phrase such as "some say." Such a phrase often but not always indicates that no one can be found for attribution. And so, back to A and B. A may have their view that not-p and there may or may not be some B who criticizes A's view. A, however, need not address B but instead may invokes a class Φ, representative of the standing opposition. Critic A attributes an exceedingly bad argument (w*) either directly to B or to Φ, responds to w*, and then claims to have defended their view. This strategy admits of some variation of attribution of w*, from a very specific B to the vaguely defined class Φ. The employment of this version of the hollow man is signaled by extremely general descriptions of one's opponents in Φ—they may be relatively specific by "Liberals," or "Republicans", and they may be very general, such as with a "some say" or "you know someone out there thinks." Some such general phrase used without naming a specific opponent often indicates a hollow man. Importantly, unlike the representational form of the straw man, this argument is not a distortion but a complete fabrication or invention.

And so this less extreme version of the hollow man can be schematized as A reviews arguments from B, C, and D, whom A classifies as the Φ's. B, C, and D give arguments aggregative to x, y, and z (and perhaps of varying quality, including x*, y*, and z*, too). Despite the rich terrain of arguments to survey and respond to, A speaks broadly of the Φ's, and instead of responding to x, y, or z (or even x*), A responds to an argument w* that has no relation whatsoever to those given by any member of Φ. Like with the case before, the problem is that A responds to arguments not given. But in this case, the argument not given is not traceable in its representation to any of the actually given arguments. It is not that A misconstrues or misrepresents some argument or other (as x* replaces x, say) but A's argument for response, w*, does not even purport to represent any of the arguments actually given. Here, the distortion is not of the arguments, because w* does not function as a representation of the arguments given (it is not about or a stand-in for x, y, z, or even x*). The argument w* is its own entity, and as a consequence, its distortion is not of any of the arguments actually given by the other side, *it is a distortion of the fact that those arguments were given at all.*

In contrast to representational straw man and weak man arguments, where the targets are saddled with less defensible versions of their expressed views, the hollow man saddles the targets with not only less defensible views and arguments but ones utterly different from the ones they have expressed. The distortion, then, is complete, because those who hollow man fail to address standing critical questions and cases against views in question, and instead proceed to box with shadows. In some sense, the term "straw man" captures the spirit of the hollow man, as what is constructed is an opponent entirely of straw, and one constructs such an opponent exclusively for the sake of making quick and overt work of it instead of anything really resembling one's actual opponent. Moreover, no one will be motivated to step in to defend it.

It is this final point that is crucial to the hollow man—namely, that because (i) the opponent is named vaguely, and (ii) the positions and arguments attributed to the opponent are not only indefensible, but also bear no resemblance to any stated commitments by opponents, no one will arrive to redress the situation by defending the opposition. That is, because A names no specific B for criticism, no one arguer in the opposition has special obligation to clarify the position or argument. And because the position bears no resemblance to any given argument, no member of the opposition is likely to recognize their case as being distorted and criticized. Moreover, because the position being criticized is one they also likely reject, it is likely that they will want to avoid addressing the issue, lest they, even in protest, risk being associated with the position.[21]

3.5 Hollow Manning in Practice

Our case that the hollow man deserves its own analysis as a form of argument depends on two factors. First is a theoretical requirement: there is significant non overlap between forms of this fallacy and those belonging to other dominant fallacy-schemata for straw man. Second is a pragmatic requirement: the special identification of the form of fallacy provides use for criticism of standing arguments and their unique implications. We have shown that, though central to the spirit of the straw man fallacy frame, hollow man arguments are significantly different from standard accounts of straw man fallacies in terms of either the representational or selectional forms. This difference, again, is that it is not a distortion of *any specific* argument an opponent has given or even their position, *but a distortion of the fact that the opponent had even given them*. It is a distortion of the overall dialectical situation by the meta-argumentative implication that the opponent has not said anything at all worth hearing, once we see what they say through the lens of the presented argument, w*. We think this is sufficient for meeting the theoretical requirement. What, the, is the practical payoff for formulating a new form of straw man fallacy? Let us begin with some examples from American political discussion.

3.5.1 Bush and Iraq

In early 2004, in the wake of the American Invasion of Iraq, US president George W. Bush was finding that the war was becoming unpopular at home. The weapons of mass destruction the CIA reported Iraq had and that precipitated the war were not to be found, Iraq was becoming unstable under coalition occupation, and many Iraqis were radicalized and consequently in full opposition to any American-supported government. Many in the United States were concerned that the situation was not hopeful for the fledgling democracy. In his State of the Union address, George W. Bush spoke to those worries with the following:

> We also hear doubts that democracy is a realistic goal in the Greater Middle East, where freedom is rare. Yet it is mistaken, and condescending, to assume that whole cultures and great religions are incompatible with liberty and self-government.
>
> (2004)

In case there was any doubt what Bush's implication in the State of the Union was, he followed the thought in a later interview in the Rose Garden:

> There's a lot of people in the world who don't believe that people whose skin color may not be the same as ours can be free and self-govern. I reject that. I reject that strongly.
>
> (2004)

Granted, it is unusual for presidents to respond to specific criticisms, but it is unclear just who George W. Bush had in mind in delivering these responses. The implication is that the arguments roughly rendered represent the main motives and reasons behind the opposition:

> Iraqi culture (or Arabic culture generally) and the local Shia and Sunni religious sects (or Islam generally) are incompatible with liberty and self-government. Therefore democracy is not a realistic goal in Iraq (or the Middle East generally).

The background inference, then, is that those concerned about the feasibility of Iraqi democracy (and by extension, American support for those efforts) are actually cultural and religious bigots. The problem, however, was that nobody in the opposition said anything like that. Surely, no serious discussant would even think to hold that populations of people darker than George Bush are incapable of democracy, as India is the world's largest democracy. Most of the concern expressed was whether institutions overtly sponsored by America could stably govern in Iraq, given anti-American sentiment. Others were skeptical of whether the Americans had given the Iraqi government enough support in either troop numbers or administrative assistance. To those in the opposition, Bush's line was positively baffling. Indeed, Edward Haley observed that Bush's strategy was to cast opposition to further occupying forces in Iraq as a "false dichotomy," so that "anyone who had even modest doubts about the democratization of Iraq was an unpatriotic bigot" (2006: 177).

The importance of Bush's casting the opposition as racist was not simply in presenting the opposition as ridiculous but as having views nobody would overtly express. Bush's strategy is to impute an unexpressed motive to those who oppose his proposals. Namely, that even if those in the opposition deny they said anything like what Bush had attributed to them, Bush has, at least in the eyes of his preferred audience, identified their real reasons. Those in Bush's preferred audience were not fazed at all by the fact that no one had said anything amounting to the expression of racism, as Jeffrey Lord explained in *The American Spectator* that the Democratic Party as a whole is driven by unexpressed racist aims:

> [T]he party of race followed its support for subjugating millions of black Americans to slavery with support for a hundred years of segregation. After abandoning millions of people of color in their struggle against Communism, it now seeks to do the same as Iraqis struggle against Islamic fascists.
>
> (2007)

Consequently, even if those in the opposition were to press the president for one Democrat expressing concerns about democracy in the Middle East on racist grounds, the strategy is to hold that even if they never said it, they nevertheless meant it.

3.5.2 Spanish Immersion

Consider a similar exchange regarding a language element in public schools in the United States. Some schools, in order to improve American students' foreign language competency, have instituted "immersion" primary (K–5) schools. In immersion schools, students are instructed in and are expected to interact in another language for half of each school day. Children in the program quickly become proficient in the specified language, and it is widely hoped that similar programs will expand into middle schools. In Tennessee, specifically, most of the immersion schools are Spanish-language. However, there are two widespread concerns about immersion programs. First is whether instruction keeps pace with correlate English instruction, as students are really learning two things during the immersed class time: Spanish and, for example, *la aritmética*. It seems unlikely that under such burdens, students can keep pace with other students learning arithmetic in their home language. Second is whether content is easily transferable between languages. So, for example, if one part of a science lesson is taught in Spanish, there is a question as to whether that knowledge will be seen as relevant (or statable in) English language science instruction later. Parents had concerns along these lines. However, when these concerns were brought to one of the school administrators at a school open-house, she responds:

> I know there is a lot of worry about Hispanics "taking over," but there's a lot to Spanish culture that we can learn from. You *can* learn in Spanish.

First, this response does not address the stated concerns, and so it fails to match a requirement of jointly weighing reasons in critical dialogue. But note, second, what the implication of not addressing those concerns but of addressing an entirely different concern is the real worries about Spanish instruction are cultural bigotry. Even if there are plausibly phrased worries expressed, they are not the root problem. Instead, they are indications of a tacit racism. Consequently, those who voiced concerns are, if they persist with questions or protest the implication, must now labor under some suspicion they must dispel. Although the administrator does not name any specific opponent, given the context of *dialogue* (as opposed to Bush's monological context in the State of the Union and the press conference statements), it seems clear that she is suggesting to parents with concerns along the initial lines that they should search their hearts and ask themselves whether these worries *really are* expressions of latent bigotry.

3.5.3 Ann Coulter on Taxes

Ann Coulter is an American conservative columnist. She opposes taxes. She, in responding to regular expressions by rich liberals that they do not want tax breaks, but *actually want to pay more taxes*, imputes a very different set of reasons to them for their commitments:

> Really vicious liberals are constantly bragging that they *love* paying taxes. They want their taxes raised even higher! The ostensible point of these boasts is to induce admiration for their deep patriotism and unbounded generosity for the poor. But the real point is to announce that they do not share the working class's

petty concern with taxes …. "I want to pay taxes" is a way of saying that, no matter how much the government takes, they will still have enough money to keep drinking Dom Perignon and making out in the hot tub.

(2002: 38–9)

At least in the literature supporting progressive taxation we are familiar with, there is no mention of champagne or hot tubs. Two things should be noted here. First, it seems that were there a class of people with such reasons, it, instead of being a reason to cut taxes, would be a reason to increase taxes. Regardless of Coulter's self-undercutting straw man argument, the second point is that the strategy implicates that whatever reasons that are given in defense of graduated taxes, the *real reasons* are those of self-aggrandizing "braggadocio" (2002: 39).

3.6 Hollow Manning as a Dialectical Device

We noted earlier that straw man arguments have an *ad hominem* abusive character to them, especially as they are deployed not just in the face of a target in an adversarial dialogue, but also in monological portrayals of those off-scene targets. They clearly are most effective when deployed not just for the sake of the target but for a broader onlooking audience. That is, with our examples above of hollow man, the opposition's character both as an honest arbiter in the dialogue and the opposition's broader moral character are called into question. In "Bush," the president's opponents not only are not giving their real reasons (and hence are insincere) but they are actually tacit racists trying to rationalize their racist attitudes with policy-wonkery. Consequently, by the argument, we may ignore their given arguments, because *we now know what their real reasons are.* The upshot is that both a *synchronic* response to a group of opponents but a *diachronic* strategy of casting their further arguments as being further expressions of their insincerity and vice. The same goes for the "Chicken Little" case, as the opponents of Trump's policies are characterized as over-reacting and so unreliable sources of political judgment.

Given that straw man argumentation, like *ad hominem*, is usually addressed to an onlooking audience, instead of deployed directly in the face of an opponent (or used as an argument for internal proof for an opponent), it is addressed to specify the features of an audience the arguer takes to be most amenable to the argument. Consequently, it is of significance what the likely conditions for and effects of accepting the various forms of straw man arguments have for audiences.

The effectiveness of straw man technique fluctuates with the audiences upon whom they are deployed. It is clear that antecedent familiarity with what the arguer's opponent had said is a factor. Further, there is empirical evidence that personal stake in the outcome can change the scrutiny of straw men arguments. Bizer, Kozak, and Holterman have noted that in testing rhetorical persuasiveness of straw man arguments, "although the technique was relatively successful among people who lacked motivation to process the message carefully, it was ineffective... or backfired... among people who had such motivation" (2009: 225). That is, the more the conclusions of a straw man argument were likely to affect the listeners, the less likely they were to be influenced by the argument, and more likely they were to be critical of the straw-manner's position. However, as direct personal significance of an argument's conclusion is reduced, the force of the argument returns.[22]

It seems right that interest in the conclusion, as a matter of determining which conclusion is correct can increase fallacy detection in audiences and yield reduced effectiveness for straw man arguments. However, interest in conclusions comes in many forms, as many discussants and audience members have interests in conclusions not simply because they want to be able to determine what outcome is best but because they take themselves already to know (or at least have a strong preference for) what outcome is right. Consequently the Bizer model is a good place to *start* with developing a form for straw man effectiveness. The fact of polarized political views on a variety of issues is unlikely to yield reliably unbiased audiences required by the Bizer model. Instead, many straw man arguments are not made with unbiased or indifferent audiences in mind but rather are made as theater for those with whom the speaker already agrees. The speaker and her implicit audience already share an understanding (or at least suspicion) of the vice and ineptitude of the opposition, and the act of responding to distorted (representational straw man), the weakest (weak man), or invented (hollow man) arguments of the other side is an act of *pseudo-engagement*. It is the illusion of having deliberated with those with whom they disagree, play-act at having done one's homework. It is a false reminder of just how right they are and how benighted (or mendacious) the opposition must be to continue their resistance. For those with an interest in the conclusion of an argument in the sense that they have strong antecedent preferences for some conclusions and suspicions of those on the other side, straw manning is the strategy of choice. And so, against this backdrop of the hermeneutics of antipathy for the target of the straw man, they have enthusiastic audiences.[23]

Take, again "Ann Coulter on Taxes." The audience for her book *Slander* is clearly other conservatives. It is certainly more clear with her later work, *How to*

Talk to a Liberal (If You Must) (2004). She is not out to change the minds of any liberals but instead to show that the failure of contemporary political discourse is "all the liberals' fault" (2002: 1). The book is a conservative's strategy handbook for political discussion with liberals. As such, because it is styled as argument analysis (its thesis is that *ad hominem* abusive is the dominant form of argument deployed by liberals on conservative critics), the book's conclusion functions as a broad characterization of the state of play on the liberal side of the issues: wrong and supported only by character assassination against critics. (We will return to this point later in Chapter 6, since fully explaining this point requires a story of meta-argumentation and the reflective tools of argument analysis as part of the construction of this fallacy—what we call *strongly meta-argumentative* errors.) Not only does Coulter's point distort the overall dialectical terrain but it retards its improvement. Those in Coulter's audience are meant to feel, regardless of the arguments the other side actually gives (they will likely rely on Coulter to tell them what liberals say), justified in ignoring them and focusing entirely on their background and unexpressed *real reasons* revealed by the hollow man argument. The same, by extension, goes for the "Spanish Immersion" audience. The administrator, though facing and actually speaking to concerned parents, was actually addressing a preferred audience of like-minded peers (whether they were present or not), revealing the "real reasons" for the concerns about pedagogical choices. If they are addressing these parents, then we must theorize how this will work on them—we will return to this in Chapter 5 with the puzzle of effectiveness and the strategies of *silencing* and *gaslighting*.

The consequence of such a strategy as hollow man specifically is that the actual arguments given by opponents, once characterized as such, are not worth attending to in any detail. In fact, once one has specified the core, "real," argument, those expressed reasons are mere window-dressing, consequences of the opposition's rationalizations, not reasons that fix any assent. This is precisely the sort of dialectical setup that leads to what Cass Sunstein has termed *group polarization* (2002). As a cognitively homogeneous group withdraws only to exchange reasons with themselves, their "argument pools" shrink, and thereby, they not only lose the capacity to recognize reasonable disputation of their views, their views progressively begin to shift to become more radical. The irony is that though hollow man arguments are addressed to those with whom one already agrees, they, over time, contribute to drastic polarizing shifts in the group's views. One can see this pattern more clearly if one analogizes success of straw man arguments to the spread of rumors. Hurtful rumors are more likely spread in populations antecedently hostile to those whom the rumor portrays negatively. For example, the rumor that the 2008 US vice-presidential candidate Sarah Palin thought Africa was a country spread quickly among liberals in America, but never caught on with Republicans (Sunstein 2009: 6, 86). The point is that straw man arguments work not just because of the depth and intensity of disagreements and their consequent ill-will but, further, straw man arguments deepen and intensify those rifts.

3.7 Appropriate Usage for Straw Man Arguments

Above we've identified three distinct forms of straw manning: the representational form, where one distorts an opponent's commitments and then holds the distorted view to critical scrutiny; a selectional form, where one selects weaker elements (or representatives) of a case and calls them to account; and finally a form where one makes up a fake opponent for the purposes criticizing them. So, the standard representational straw man, the selectional weak man, and the hollow man. While our account to this point has clearly expanded what counts as a pernicious misrepresentation (and added new metalogical terminology to boot), we think this is really only part of the story of the straw man. To continue this story, and to close this chapter, we would like to consider a puzzle: *Is it ever permissible to straw man*? This question arises from the now standard tendency in argumentation studies to note that many fallacy forms have legitimate usage in some contexts. For example, circumstantial *ad hominem* is formally identical to showing that a purported authority has a conflict of interest as to what they say. It is only that some conflicts are relevant (e.g., that one's research was funded by someone who prefers some outcome) and some are irrelevant (e.g., that one is a church employee and also a defender of its practices). Current accounts of fallacies are incomplete if they don't consider whether there are non-fallacious instances. In the case of the various forms of straw manning, however, it seems completely counterintuitive that one might ever do it legitimately. Christopher Tindale notes:

> A "Straw Man" argument would seem to be always incorrect and have no redeemable instances. This means that we cannot define "fallacy" as the misuse of a legitimate argument strategy because… there are recognized fallacies that do not fit it.
>
> (2007: 12)

It seems, on its face, that Tindale is right—the frame for straw man arguments and their function over time can never be legitimate. For, at the very least, one would be endorsing a false premise and truth is the basis of all cogent argument. But this doesn't appear to be the always the case. Jan Albert van Laar (2008) has proposed that there are contributions to critical discussions that are formally similar to cases of straw man, but are dialectically reasonable:

> Critics have some room for maneuvering when raising critical doubt, even when they change the arguer's position or formulation, because there are contexts of utterance within *confrontations* where such a contribution does not allow for a *reconstruction* in such a way that the result is not a *direct* and *fallacious* controversial *maneuver* but, rather, an indirect and dialectically reasonable one.
> (2008: 196)

On the strategic maneuvering model of critical dialogue, it must be assumed that each party in a dispute "strike a balance between the shared dialectical objective of dispute resolution and his or her individual objective to persuade the opponent" (2008: 198). In turn, there must be room for parties to reconstruct opponents' arguments in ways that emphasize coming criticism, direct further critical discussion, or raise background concerns. In instances of critical exchange, Lewiński terms this "the strategy of easiest objection" (2011: 492). On this reasoning, van Laar also holds that there must be "leeway" for parties to (i) "find a difference of opinion they both find interesting and worth discussion," (ii) "work toward their individual rhetorical aims," and (iii) "attempt to get at the real, underlying position of an arguer" (2008: 204) by representing and critically responding to each other's standpoints. The critically advantageous reconstructions are allowable only if they meet the following conditions:

1. The critic makes it clear to the arguer that the standpoint criticized is one that is reconstructed (and thereby requests clarification, if it is excessively distorted).
2. The reconstructed standpoint must not be one, given the context of the standpoint's expression, that would not be clearly rejected by the defending party.
3. The reconstruction contributes to further (and improved) critical discussion of the issue (from van Laar 2008: 204–6).

Under these admittedly restrictive conditions, it seems that what might seem straw man arguments of any of the three forms can actually be reasonable challenges in critical dialogue.

That is a surprising result, for sure, but consider the following examples.

3.7.1 Religion and Morality

Albert: Religion has had such a distorting effect on morality. Look at the way that inter-religious wars and suspicion have made people blind to human suffering.

Betty: But look, not all religion has had only that effect. For example, look at golden rule ethics. That's on the right track, right?

3.7.2 Moral Atheism

Xavier: You can't be moral without religion. That's why atheism is wrong.

Yelena: I don't see how atheists can really be moral. Religion is the source of morality, so you just have to believe in God.

Zed: It is possible to be moral without religion. But God exists, so we don't need to worry about that, now.

Adam: There's a regular thought that one cannot be moral without religious belief. Here is how it is possible…

3.7.3 Aunt Mary

Penelope: Let's skip visiting Aunt Mary today. It's such a long drive, and gas is expensive. And I'm not feeling well. And her house smells funny. And her dog bugs me.

Sam: Now, you're just saying those things because you don't want to miss your television program. That's not a reason to skip seeing your Aunt.

Let us take the examples in turn. In "Religion and Morality," it is unclear what Albert's stated position is. It could be that all religions always have distorting effects. But that's not clear, as he could also mean that most forms of religion tend to have the effect. It is not clear how strongly Albert takes the connection between religion and moral distortion to be. Betty's challenge imputes, by her reason given, a strong and implausible interpretation of Albert's commitment as being quantified universally. She implicates this by the fact that she presents golden rule ethics as a counterexample. Betty has thereby saddled Albert with a less defensible view than he expressed (albeit vaguely). But her move, though structurally identical to a representational straw man, is positively helpful to the conversation, as Betty's challenge allows Albert to focus on how to qualify his claim. It may, further, direct how the two discuss whether golden rule ethics themselves are consequences of religious views or actually what may be distorted. The point, of course, is that Betty, in imputing a less defensible claim to Albert, instead of distorting the overall dialectical terrain, actually has contributed to clarifying it. Douglas Walton has rightly noted that straw man arguments (in what we've called the representational form) generally function in the vagueness of what exactly a speaker's commitments are (1996: 125). They, on the one hand, may fallaciously take advantage of how poorly defined those commitments are. However, it is clear here that the argument form can also be used non-abusively to open the door to clarify and precisify those commitments.

In "Moral Atheism," Adam responds to Xavier and Yelena's arguments, but he ignores Zed's. Zed's is likely the most advanced of the arguents, and Xavier and Yelena's don't have the metaphysics on their side (as the divine command theory of ethics seems wildly implausible). But Adam is right to address them most explicitly at the beginning, and perhaps exclusively so. Adam's dialectical strategy is to address his opponents in a way that is most responsive to the broadest commitments of the audience. He takes the temperature of the dialectical community on an issue, and he first addresses the arguments to what he takes to be the most widely held misconceptions about atheism. Many of these misconceptions are based on simple errors (e.g., an uncritical acceptance of divine command theory), and before any more advanced discussion on an issue can proceed, those basic points must be set aright. And so, even if a weak argument x* is not the best argument given by the opposition, it is important to address it, especially if it is widespread and easy to mistake one's way into, and its prevalence would inhibit further discussion of other arguments (x, y, or z). As Brian Ribeiro (2008) has noted, strategies like weak manning a viewpoint are sometimes necessary for pedagogy in an area, especially given the many non-argumentative constraints on critical dialogue:

> Some less-than-stellar versions of p may be of *considerable historical interest*... or perhaps the best available version of p is just *too hard* for our students... or... sometimes we even give some weight to the overall ease of *delivering the course materials* to the students in an accessible format. So I teach *these* defenses of epistemic externalism... because *these* are the ones in the... anthology I use.
>
> (2008: 31)

Ribeiro holds that these selections are nevertheless distortive but are ineliminable, given the context. There are, of course, ways to mitigate the concerns, and one, of course, is to frame the responses (and even where the discussion ends) as propaedeutic for further discussion. The problem is that critical discussions must *start somewhere*, and especially when the context is fraught with many bad arguments and crazy views, it is best to start by addressing them, else they will derail more developed discussion. The problem is that discussing them can overshadow the more nuanced and dialectically robust critical discussions. The crucial thing, then, is for both arguers and their audiences to be careful not to make the hasty move of casting opposition in the form of the most common errors made by its prominently criticized representatives.

Finally, "Aunt Mary" is a case where Penelope *is* rationalizing, or at least, it is appropriate for Sam to challenge her on whether she actually is operating on other reasons than those given. Addressing her given arguments, until this concern is addressed, would be a waste of Sam's time and breath. He must speak to her *real reasons*, even though she may not express them. And so Sam may rightly ignore Pen's rationalizations and speak to the reasons that may actually resolve the issue. What makes it appropriate for Sam to proceed as such is that he has either correctly or at least plausibly diagnosed the dialectical situation and has proceeded in a way that allows discussion of the real reasons Penelope wants to skip the trip. Surely, Sam must labor under a dialectical burden of his own. Roughly, he must show that Penelope has not been forthright with her reasons, perhaps by showing that the things she's protesting now haven't yielded resistance on other occasions, but only show now that the TV show is in the mix. What is important, though, is that Sam's contributions are with the objective of allowing all the reasons bearing on the decision are brought to critical light.

Jennifer Nagel (2019) calls similar strategies of clarifying or requesting further information "fishing" conversational tactics. The rule in the background is that if one conversant makes claims about the things the other is jointly acknowledged to be the epistemically better off about, the claimant is best interpreted as asking for information. So, if in the midst of your story about your rainy camping trip, your neighbor says, "and then you drove into town and got a motel!" she is not saying that she thinks you did that. She's making a joke and asking you to then tell her about how you hung your tarp and ate cold beans in a puddle—but loved it. Straw manning can take this kind of fishing form, too. A speaker can restate a straw man target's views back to them in a less plausible way as a way of allowing the target to clarify the view. It is a form of inquiry about the conversant's view, reasons, and broader account.

We have seen, then, that, structurally, straw man arguments can be non-fallacious. Representational straw man can serve to focus discussion or be a form of argument that a standpoint needs clarification, as seen in "Religion and Morality." Selectional straw man (weak man) can function as a strategy of clearing the air or correcting common errors, and thereby serve the role of allowing more robust argument later, as seen in "Moral Atheism." And finally, hollow man can be used as a diagnostic tool for redirecting critical discussion to reasons that actually determine the assent by some discussants, as seen in "Aunt Mary." What makes straw man arguments fallacious, from a dialectical perspective, then, is not their *synchronic form* of distortion or misrepresentation but their *diachronic function*. The functional problems with straw man have widely been recognized. Walton, for example, is careful to note that the formal misrepresentation is not the core problem but the way it functions over time:

> The straw man fallacy is not simply the misrepresentation of someone's position, but the use of that misrepresentation to refute or criticize that person's argument in a context of disputation.
>
> (1996: 124)

However, a finer edge is necessary in light of the examples above. We will turn to this in the next chapters. For now, we can say that it is not the *critical* function of straw manning that constitutes its fallaciousness, as similar critical distortions have been shown to be helpful. Rather it is the straw man's meta-argumentative *closing function*, namely, that of portraying the discussion critically over with the opposition, in light of the arguments given and criticized. Arguments function fallaciously when they impede the rational resolution of an issue. Representational straw man is fallacious when the weaker view's refutation does not clarify the issue, but only is made to score rhetorical points on an opponent (and thereby, impugn their intellectual character). Weak man arguments are fallacious when they present the opponent's weakest arguments as representative of the arguments given even by the best of the opposition. Their refutation then closes the issue prematurely. Hollow man arguments are fallacious when used as evasions of the burden of proof to respond to the standing arguments given by an audience but instead, to address arguments not given. Instead of *adding to* the reasons under scrutiny, they only *distract*.

To sum up: we have argued here that the straw man fallacy is actually a family of related forms of misrepresentation of the opposition in a critical dialogue: the representational straw man, the selectional straw man (weak man), and the hollow man. Each of these forms is distinct, yet all, when fallacious, function as impediments to further critical engagement with opponents. However, there are formally similar maneuvers in dialogue that contribute positively to rational resolution of a dispute. Consequently, what makes straw man fallacies fallacious is not simply the distortion of stated views of opponents but the use of these distortions to close further critical engagement on an issue.

4

Straw Men and Iron Men

4.1 Introduction

The brief history of the straw man we sketched in Chapter 2 showed that the straw man evolved from Aristotle's *ignoratio elenchi*, when it was taken, mistakenly, to mean "irrelevance of the proof." An entire tradition of interpretation was then constructed around this notion of relevance. With the advance of years, and the proliferation of textbooks, new varieties of irrelevance were discovered and included in the general account—*ad hominem*, *ad fidem*, *ad baculum*, *ad misericordiam*, and so on. Eventually, *ignoratio elenchi* was left to cover remaining as-of-yet unidentified or unnamed cases of irrelevance. But, as we've seen, relevance, or irrelevance, is an imperfect way of characterizing the straw man. In fact, showing that a bad view on the issue at hand is false or that a flawed argument bearing on the debate is incorrect *is* relevant. That's why the straw man has a core of correct and even, on occasion, useful contributions to critical exchange in non-fallacious instances. While it's true that a modified case is irrelevant in some sense—because, in some cases, it's not one's interlocutor's view that's at issue—the primary feature of straw manning is misrepresentation of the dialectical terrain and then on the basis of that misrepresentation, prematurely closing an issue.

Chapter 3 then presented an expanded basic picture of the straw man fallacy. Rather than one basic form, the straw man of the twentieth-century textbook tradition, there are at least three: the *representational* form, or standard straw man, distorts an actual interlocutor's position so as to more easily defeat it; the *selectional* form, or weak man, accurately critiques a weak facet of an opposing case but misrepresents its salience to the overall argument; finally, the *hollow* man consists in fabricating an opposing view and defeating it. The expanded forms show that the basic dialectical picture of argument is more complex than it might otherwise first appear. This complexity arises from a core dialectical feature of reasoning, which requires us to construct an image of the available alternatives. All arguing or reasoning, we are maintaining, paints a picture of an alternative that is wrong or less well supported than the option supported. In this chapter we follow the thought that the phenomenon of straw manning alerts us to the richness of dialectical obligations. These are not exhausted, so we have argued to this point, merely by recommending, as much in the tradition has done, scrupulosity with regard to the characterization of a specific argument at a specific time (which in itself in the case of the weak man might be a means of dialectical distortion).

Our account of the straw man to this point captures the traces of it present in the literature dating back to Isaac Watts (1725). Even though our story is more elaborate, to this point it is still a basically traditional account in the sense that our picture of the straw man has it that *negative* misrepresentations of an *opponent's* view are very often, but not always, dialectically pernicious. In other words, a straw man is deployed in the context of a debate against a hapless target. We have discussed the ways this can happen and we've identified some limiting criteria. But, as we shall argue here, straw manning is not limited to negatively adversarial contexts. Misrepresentations serve several epistemically significant ends. But this also points us in other direction. One of the central worries about sophistry, as Socrates notices in the *Apology*, is that bad arguments might seem to be good ones. So we now ask: What about representations that *improve* an argument? Are they fallacious? We think they can be, and when they are, they are *iron men*.

The iron man presents us with a puzzle. The puzzle arises from the long-standing advice on how to avoid straw manning: be charitable; stick to the point at issue, and don't wander; and don't let winning get in the way of a good argument. If anything, then, the iron man represents the core virtues of arguing. But, alas, it seems that all good argument advice has a dark side. For charity, there is toxic charity; sticking to the point makes us warp the dialectical context and miss salient questions from unanticipated perspectives, and not being competitive enough allows bad arguments to go unchallenged. The distorting effects of this are as pernicious as the classic straw man. Argument, after all, is meant to clarify the quality of reasons, and therefore the quality of reasoners. The iron man, when performed inappropriately, stands in the way of this. Strangely (is it strange?), the iron man is a problem of insufficient adversariality in argument.

In the last chapter we discussed the now common view in argumentation theory that fallacious arguments are deviations from otherwise legitimate argument schemes (see Tindale 2007; Walton 1999; Walton, Reed and Macagno 2008). As we discussed above, this might not apply in the case of the straw man. As noted in the previous chapter, Christopher Tindale argues the straw man is never correct (2007: 12). Indeed, the notion of distorting someone's argument for some legitimate purpose seems counterintuitive. While *ad hominem* arguments may be legitimately employed to undermine an arguer's credibility (when that credibility is relevant), distortions of an interlocutor's argument seem to have no place. However, we've shown that there are exceptions to this general rule, so there can be argumentatively salutary straw man arguments. Developing this idea, we think, will pay significant theoretical dividends.

A second, and more profound, challenge to the view that no legitimate straw man schemes exist consists in denying that it is an identifiable scheme at all. Douglas Walton and Fabrizio Macagno write:

> With many of the informal fallacies, the problem of fallacy identification, analysis, and evaluation is made easier by the fact that the fallacy is closely related to a known argumentation scheme (Walton 1995). For example, the fallacy of improper appeal to authority is based on, and can be evaluated, using the argumentation scheme for argument from expert opinion. This resource is not available, however, in the cases of wrenching from context and straw man. These fallacies are more purely dialectical in nature.
>
> (Walton and Macagno 2010: 303–4)

Straw men arguments involve at least two arguers, with one representing and attacking the argument of another, the target. This attack, as Walton and Macagno point out, may take the form of any argument. In fact, for this reason, it would be more apt to consider the straw man a dialectical operation, tactic, or ploy rather than an argument scheme. Indeed, that the straw man lacks a specifiable scheme (unlike, say, the *ad hominem* or the *ad verecundiam*), yet remains persistent wherever arguments are made, means that another approach to the phenomenon is necessary, one that captures the richness of its employment and clearly delineates its fallacious from its non-fallacious use. We believe we have found such an approach in the virtues of argumentation, in particular, the virtue of open-mindedness. The key to all of this is that open-mindedness is a meta-cognitive virtue—one that is having attitudes about other attitudes held by others, willingness to hear them out and be interested in their grounds. Straw man fallacies are meta-argumentative fallacies, as they are fallacies about arguments, so they are rooted in attitudes we take about others' reasoning. Meta-cognitive virtues, we hold, are part of how we address and manage straw manning, and much of those virtues are dependent on broader contexts of argumentative trust and regard, as we will show with some of our salutary cases of straw man arguments.

In addition to capturing what makes standard instances of straw manning fallacious, this approach has two further reasons to recommend it: (1) it shows when straw manning is not fallacious (and so answers Tindale's objection) and adds more nuance to the previous examples, and (2) it uncovers a little noticed variety of straw man—the distortion which results in being overly charitable to someone's argument, or, as we have called it, the iron man. Fallacious varieties of straw man, we shall argue, arise from vices of closed- and excessively open-mindedness: being too critical in the case of straw manning and being not critical enough in the case of iron manning. Importantly, the virtue approach shows how an argument might be a straw man in one context but not in another, as, again, virtues of open-mindedness are practicable in conditions amenable to those virtues. Environments without dialectical trust and regard make the virtues more difficult to be effective. After all, in true Aristotelian fashion, virtuous arguers aim for the mean with an eye toward offering the right argument, to the right person, in the right way, at the right time, and in the right context.

4.2 There Are Legitimate Uses of the Straw Man

The various schemes of straw men are defined by the way one arguer represents the views of another: badly, selectively, or falsely. We saw in Chapter 3 that it seems there are ways that one can badly, selectively, or falsely represent someone's views without being guilty of fallacy. Now let's take a closer look.

Let's begin our account with a thought experiment. Imagine teaching a philosophy class where every argument had to be presented in its most accurate form with no substitutions. Students therefore would have to cut their teeth on the real version of Anselm's Ontological Argument. It doesn't take much reflection to realize this would be a nightmare; in fact, it would be very hard to teach philosophy without employing some variation on the straw man scheme frequently and energetically. As we've noted in Chapter 1, this is especially the case in fallacy pedagogy and theory. With regard to this reason, we've already mentioned that Brian Ribeiro (2008) has argued that that distortions formally identical to straw man distortions occur frequently in the classroom from pedagogical needs of historical interest, pedagogical ease, and practical availability. There seems, in fact, to be an intuitive case for using the various schemes of the straw man pedagogically. Representational straw men might be employed to drive home particular pedagogical points. In philosophy, it's a regular enough phenomenon, as most views worth consideration have well-developed and nuanced versions as the going, best representations, but they are inappropriate for presentation to the average sophomore. So, to start, more rudimentary versions can be presented, if only to show the stakes and whet the students' appetite for the critical dialogue to come. All this is stage-setting for the importance of the nuanced versions, if they are to investigate them. So for example, when fatalism is taught, the view that all truths, insofar as they are true, cannot be otherwise, simple theological versions are the best place to start. So a philosophy professor may start as follows.

4.2.1 Fatalist Prof

> Professor Melissius: According to a venerable tradition, fatalism is true, if only because it is a consequence of God's foreknowledge and creation of the world. So when God chose to make this world, he chose all the things that make it *this* world, including all the events of the past and the future. That things could have been or could be otherwise is an illusion created by our ignorance of why they had to happen.

The problem with theological fatalism, of course, is that it depends on a particular theology being defensible, and it comes at the cost of a puzzle over whether it is consistent with doctrines of punishment for sins, as they seem to require freedom to do otherwise as a precondition. Moreover, there are questions of whether the choice of the deity to make this world instead of another implicates that things could in some minimal sense be otherwise—God could have chosen differently, but just didn't because this is the *best* world (which, as a comparative, implies other possibles). Metaphysics is difficult business, but in light of all these questions, starting with theological fatalism, is nevertheless a good pedagogical choice, because selecting the weaker version of this view makes for quality instructional discussion with students. They get the thrill of scoring a point on a view, and teachers get to show them how a view can be improved while under the fire of critics. A final reason in favor of starting with the theological version of fatalism is that though it is the least defensible version, in the history of philosophy, it was the most widely held version of the view, starting with Plato through the Stoics and into the modern period with Spinoza. It makes sense to teach it, if only because it's the version so many of the biggies hold, even if they could do better with causal or logical versions of the view (which they arguably held, too, but those require that we squint pretty hard). Of course, if students continue to push on the matter, these more defensible and argumentatively complicated versions of the view await inquiring minds, but it is because we start by weak manning fatalism with its initial and simple theological version. Consider another case, but this time, one that straw mans the student's view.

4.2.2 Hamlet's Ghost

> Professor Elba: Hamlet has no stage directions for the Ghost. None. I don't think he should be onstage for the production.
> Lydia: But, look, professor. Shakespeare gives very little if any stage direction for any of the characters in any of the plays. Surely Julius Caesar should be on stage, but there aren't any directions for him.
> Professor Elba: But what's the idea for staging the Ghost? Having somebody stomp around stage with a bed sheet over his head?

Of course, the idea of the Ghost in *Hamlet* being dressed in a white bed sheet with eyeholes cut out would be laughable, but Professor Elba is being coy with Lydia. When Professor Elba proposes the absurd view to represent Lydia's position, she is offering Lydia an opportunity to improve the thesis. Professor Elba is providing an occasion for her student to read a text with an eye to the breadth of responsible interpretation and staging. When Professor Elba proposes the absurd bed sheet view for the Ghost-on-stage position, she isn't closing off the argument, but rather, she's setting parameters for fruitful discussion.[1]

A similar case might be made for the other two straw man ploys. A weak man might be used as practice.

4.2.3 Gay Marriage

> Brad: I've heard quite a number of arguments against gay marriage in the conservative press lately.
> Angelina: I have too. I heard one particularly bad one from a speechwriter (and blogger at RedState.com). He wrote, "It does not affect your daily life very much if your neighbor marries a box turtle. But that does not mean it is right …. Now you must raise your children up in a world where that union of man and box turtle is on the same legal footing as man and wife."[2]
> Brad: Wow, that's hilarious.

In this example, Brad signals that there are several arguments against gay marriage. We can imagine that some are better than others. Angelina responds by attacking what is likely to be weakest of them, a kind of textbook version of the slippery slope fallacy. Answering it first improves further discussion. We might call it a *clearing the decks* function or FAQ procedure—one must call out and address widespread bad arguments before one can turn to the good ones. So long as the bad arguments are addressed up front, the discussion has the chance to progressively improve.

For a hollow man case, we must extend our pedagogical considerations. As we noted in the Introduction, there are practical, political, pedagogical, and philosophical reasons for using straw manned examples in the teaching of argument. As we noted in Chapter 1, open just about any introductory logic text, and you will find the exercise sections full of arguments few sensible people would make (though social media has made us question this notion). Despite the deep psychological costs of coming up with a series of awful hollow-manned arguments, it's just easier and more effective to teach the fallacies this way, for the point of the fallacy exercise is to get at the form of argument, not to pin failings on specific people. Here is a representative sample from textbooks simply snatched off of Aikin's bookshelf at random.

4.2.4 Textbook Fallacies

I'll tell you why I believe this war is right—I love my country!

Look, you know socialism is wrong. What would your friends say if they heard you deny that?

The theory of evolution boils down to the idea that your great-great-uncle is a monkey.

If we recognize Cuba diplomatically, then we will end up granting that same status to every country run by an anti-American dictator.

The point, again, is that we have such hollow man versions of these fallacies as textbook examples not because they are representative of how people regularly argue but so that our students can master concepts of criticism with easy and manifestly bad cases. They, after having these large, slow-moving targets for criticism filled full of their critical arrows, can later be able to manage these concepts in harder more ambiguous and arguable cases. We would like to stress that contextualizing these hollow man cases is the job of the conscientious and diligent instructor. Mixing these up for the real thing is like a boxer expecting his opponent will be a literal punching bag.

Though all of these examples fit the straw man ploy in its various forms, none of them are in our view fallacious. In "Fatalist Prof," Melissius introduces the theological version of the view for the purposes of inducing student understanding and also the well-run criticism of a view on offer. That's just philosophy—not just understanding big ideas but giving them the business. The instruction need not end with the criticisms surveyed, but can move on as refinements to the fatalist view or its alternatives are given air time. In "Hamlet's Ghost," Professor Elba straw mans her student's view in order to spur her to improve it. In "Gay Marriage," Angelina goes straight for the weakest of the arguments for the anti-gay marriage position, and so weak mans that view. But she does not draw the inference that the criticized argument or view is representative of the best of the opposition. Weak manning sometimes serves the dialectical purpose of clearing away weak arguments, what we called *clearing the decks* in Chapter 3, which nonetheless may have many adherents, and which nonetheless occupy much in-demand dialectical space. And sometimes, we might simply introduce hollow men for target practice so that we can master the concepts of criticism, as we see in "Textbook Fallacies." They don't stand for any going version of a view, but they sharpen our critical vocabularies.

These representative, but non-fallacious, straw man tactics highlight two important features about what makes most straw man arguments fallacious in the first place. The fallaciousness does not primarily consist in the distortion of someone else's argument (as in the representational straw man), in the distorting selection of the weakest of someone's arguments (as in the weak man), or finally in the invention of weak arguments or arguers (as in the hollow man); all of these can be very useful dialectical tools. What makes these tactics fallacious *is how they are deployed*. That is, they are deployed meta-argumentatively along the following lines: we see with this track record of bad arguments from this target group or individual that the case for the other side on this matter is particularly weak, and the quality of these arguments bespeaks unserious minds that continue to endorse and give them. This meta-argumentative inference, one based on assessing arguments and their arguers, yields a conclusion about the future for worthwhile argument on this matter: it is closed. This, we've called the *meta-argumentative closing function* of straw man arguments, and it is abused in fallacious straw man cases.

Another way to see this point here about impugning interlocutors and closing arguments is to see how what we'd called the non-fallacious or salutary straw man arguments worked. Most of our examples here are pedagogical, but it's worth noting that it might take well-regulated spaces for these exchanges to work as they do. But "Fatalist Prof" and "Hamlet's Ghost" function as instances where a view is straw manned for the sake of opening a wider-ranging discussion. "Gay Marriage" is taken on for clearing the decks for a more appropriately focused take on arguments. And so long as "Textbook Fallacies" functions as presented as target practice, slow-moving but unrealistic opponents, nothing is impeded for any arguments with the ideas on offer for them. And note that from Chapter 3, our example of the hollow man with "Aunt Mary," where the submerged real reasons are addressed, not the rationalizations given, we have genuine progress on the actual reasons that yielded the dispute. In every case, the chances for continued reason exchange are promoted, and either the opponents are rendered as convenient fictions or as ones with things to say back to the critics. But in every case we've surveyed as salutary, a background assumption has been in place, and it is that even if the argumentative exchange is minimally adversarial, there is sufficient trust in the context for the straw man to spur further conversation and revision. This is especially so in pedagogical contexts. As any experienced teacher can attest, a loss of goodwill in a classroom undercuts many otherwise effective pedagogical and argumentative strategies. In "Fatalist Prof," with Melissius opening the discussion with a weak version of a view, if the student audience mistrusts Melissius with the discussion, disaster is likely. For example, consider an analogous case.

4.2.5 Fallibilist Prof

> Professor Tullia: If skepticism is false, we know many things. But if that's true, then we must know on the basis of incomplete reasons.
> Renee: But how can one really know with incomplete reasons?
> Tullia: Ah! Yes, it seems if we are fallibilists, we must be able to say that we know, but may still be wrong.
> Renee: But if you know, you're not wrong.
> Tullia: Right.
> Renee: You are now contradicting yourself.

The problem for Professor Tullia is that she opened the discussion with a paradoxical statement of fallibilism, a common one, but one that yields problems quickly for fallibilists who hold that knowledge is possible on the basis of incomplete reasons—everything depends on what "complete" would mean. Professor Tullia's objective is to draw Renee into the discussion with the straw man of fallibilism and its appealing anti-skeptical results but curious view about reasons. Tullia hopes that Renee will be keen to participate in the view's revision as the difficulty arises. But if Renee is inclined to think that Tullia's imperfect first statement is the considered view, it follows that revisions will look like argumentative incompetence—Tullia keeps, from this perspective, contradicting herself. The development of a view, from the take of an audience not sympathetic with the process, appears like the speaker can't keep her story straight. So particular argumentative virtues on display with these argumentative tactics (with both speaker and listener) depend for their effectiveness on whether the context is one open to such moves. Some relationships have developed to the point where there is trust, or some contexts are one where there is antecedent tolerance for the longer way to instruction. However, some relationships have soured to the point where those communicative tactics will backfire.

A similar point can be made with "Hamlet's Ghost" and "Aunt Mary." If the two conversants do not have either established relations of intellectual trust, mutual regard, or minimal cooperative attitudes, the strategies may be more occasions for argumentative escalation. Professor Elba in "Hamlet's Ghost" certainly takes a risk with Lydia, as such an interpretive move could be highly alienating if done in the wrong spirit or even with the wrong tone of voice. And the same goes for the "Aunt Mary" case, as the challenge there is that there are hidden and untoward motives. So what we've called the closing function of the argument is highlighted if the parties are not interested in ongoing conversation. And it is worth noting that this strategy of using straw man strategies as argumentative hooks is epistemetrically costly, as it is an indirect method for instruction and clarification. Consider that communication is costly in time and intellectual energy, so it is usually the best use of resources only to pay attention to high-quality reasons and alternatives. So swatting down straw opponents, even if it is clear that they are of straw and for moving things along to higher quality further items for consideration, could be taken as an inappropriate use of one's time and energy and that of one's conversant, too. So the interpersonal conditions for the salutary instances of straw man arguments must be in place for them, regardless of how they may be posed to the target speaker, to not have derailing effects on the dialogue.

4.3 The Iron Man

If what makes the varieties of straw men fallacious is their meta-argumentative exclusionary, or closing, function, then it is easier to distinguish fallacious cases of straw manning from non-fallacious ones.

The fallaciousness of straw man arguments is indexed to context. Views or arguments that warrant careful consideration in one situation may not deserve them in another. This means at times it may be permissible (and necessary) to exclude some views from consideration on the basis of cursory arguments. In other words, while fallacious straw men involve the exclusion of arguments or arguers from justly deserved consideration, in light of the function of the straw man to distort over time, there is good reason to think that unreasonably or overly charitable interpretations of arguments (or arguers) can also qualify as fallacious. It's certainly fallacious, in other words, to distort a person's argument in order more easily to knock it down (and malign the person as a competent arguer); however, by parity of reasoning, a charitable distortion to present an unserious arguer as serious is equally problematic. We call this the problem of the *iron man*. Consider the following cases.

4.3.1 Eric Cantor

Eric Cantor was the Republican Majority Whip in the US House of Representatives. In an interview with CBS's 60 Minutes (1/1/2012), Correspondent Leslie Stahl asked Cantor to square the fact that Ronald Reagan raised taxes during a recession with the then current Republican Party view—allegedly inspired by Reagan—that taxes ought *never* to be raised. In response, Cantor denied that Reagan ever raised taxes. His spokesperson interrupted the interview, alleging that Stahl did not have her facts straight. But she, in fact, did. Coming to Cantor's defense, one blogger (Jim Hoft) made the following claim:

> Stahl, was not being honest. When Ronald Reagan took office, the top individual tax rate was 70 percent and by 1986 it was down to only 28 percent. All Americans received at least a 30 percent tax rate cut. Democrats like to play with the numbers to pretend that Reagans [*sic*] tax increases equalled [*sic*] his tax cuts. Of course, this is absurd ....
>
> Unfortunately, Steve Benen at the *Washington Monthly* continued to misrepresent Reagan's record on tax cuts. It's just soooo difficult for liberals to understand that tax cuts work. Sad.

Notice that Hoft has offered a different (and much more defensible) view on behalf of Cantor: on aggregate, taxes were lower after Reagan's years in office than before. This was not the point under consideration. The net effect of this is to distort the proper evaluation of Cantor's claim and Stahl's criticism.

Let's call this *the problem of dialectical scorekeeping*. The problem is that Hoft, in introducing a higher-quality view for consideration, has attributed it to Cantor, instead of noting that everybody in the conversation up for consideration dropped the ball with the complex point. That's worth noting, and would be a genuine improvement to the conversation, one that adds nuance to the Reagan question and to how contemporary Republicans can look at taxation generally. That would be a salutary contribution. But by attributing the improved aggregative view to Cantor, Hoft has iron manned him. This may have been an improvement to the conversation, had he not, in iron manning Cantor, straw manned Stahl. She was not criticizing the aggregative case posed (and attributed by Hoft) but the claim that Reagan never raised taxes. And note that the function of straw manning of having not only a misrepresentation of an opponent's view on record is not the only matter here, as Hoft generalizes with his closing meta-argument not just to Stahl's intellectual character from this incident but to that of liberals generally. He concludes that it is "soooo difficult for liberals to understand how tax cuts work." Our point here is that iron manning can have a distorting effect on the dialectical score, and thereby misportray the contributions of critics of views receiving the iron man treatment. This has bad downstream, argument-closing, effects.

4.4.1 Westboro Baptist Church

In the late 2000s and the early 2010s the Westboro Baptist Church was known for demonstrating at the funerals of fallen soldiers. At their protests, they held up signs alleging that the death of the person is God's punishment for the tolerance of homosexuality in the United States. They expanded to protesting any event associated with LGBTQ rights. In light of this, consider the following exchange.

> Sally: The Westboro Baptist Church picketed my local synagogue, carrying signs that say "God hates fags." Their views are patently ridiculous; far from even the fringe of conservative Christianity. People should just ignore them.
> Priscilla: Yes, but aren't they really suggesting that our fate as a nation is bound up with the moral fiber of the American people? As we lose our sense of commitment, steadfastness, and courage, we will not realize our plans.

Priscilla raises some interesting points, but they are vaguely related to the actual content of the Westboro Church's protests and Sally's objection. The question is whether these particular arguments from the Westboroites deserve consideration. Indeed, iron manning can be an occasion for broader discussion; but one iron mans precisely to avoid the defects of the particular argument before them. But sometimes one must take out the trash.

It is worth pausing to appreciate what Priscilla was up to in this case. She extended a broader conservative social philosophy to the Westboroites—one emphasizing a shared moral core for fellow citizens to be mutually intelligible, for us to see ourselves as on the same team (if on a team at all). This communitarian critique of liberal programs of protecting individual rights has taken the form of a criticism of liberal individualism generally. Religious leaders regularly invoke this line in their criticism of the arc of US history. And while it is certainly possible that this classical Burke-style conservative line may be behind the scenes, the rhetoric of the Westboroites is more appropriately read as simply invoking the notion of a wrathful Bronze-Age god insisting on obedience with regard to prohibitions on sexual behavior. Certainly Priscilla's considerations deserve long, thoughtful, reflection, but if they are to be answered or conceded, it is not clear how that would change the evaluation and reception of the case coming from Westboro.

The point is that attempts to improve the quality of topics for consideration can have a salutary effect, but the changes in dialectical score distort what the results of the critical discussion will be.

4.4.2 Philosophy Student I

We have discussed above how teaching philosophy to undergraduates often depends on strategically employed, non-fallacious straw men. As it is necessary sometimes to straw man views, it is also necessary to iron man the student's view. With this in mind, imagine the following teacher-student exchange.

> Alfredo: Rawls' "Original Position" seems impossible to me. I mean, how are we to know what sorts of things we'll be interested in if we don't know anything about ourselves?
>
> Professor Zoccolo: That's an interesting point, Alfredo, you're suggesting that Rawls's Original Position does not take cognizance of how we are constituted by our social relations. Thinking them through abstractly seems problematic.

Alfredo's view certainly trends communitarian, but it would be a stretch to suggest that this is what he meant. Unlike the previous cases, however, Professor Z's iron manning Alfredo shows him how to improve his contributions to the discussion.

It is a common-enough occurrence in a classroom discussion that a student's view tracks with a well-developed line of thought, and putting them into the position of representing those programs can salutary consequences. And it yields an easy on-ramp for students to the broader discipline. They feel connected to the debates, feel the stakes behind the various sides and why they lined up as they did. And they have exempla for how to improve their views. Just as there can be ethical exempla such as saints, civil rights leaders, and innovators, there can be intellectual exempla, those whose views and intellectual character can inspire us. And by seeing a little of them in us, if only by reflection in a well-framed iron man, we edify ourselves and our students.

4.4.3 Philosophy Student II

The norm of iron manning student views can yield good results. It shows students how to improve their thoughts. However, it can yield classroom disaster, as it can encourage more poorly stated views. Iron manning the student makes it such that the teacher does the work in crafting the views. Moreover, time in the classroom is too short to take all the off-the-wall views seriously. Sometimes, iron manning undercuts a serious classroom discussion. Consider:

> Professor Barleycorn: Descartes' argument in the First Meditation is that very little of what we take ourselves to know securely is certain. It may all be a dream. Or it may all be an illusion of a very powerful demon.
> Bradley: Dude! I had a dream like that one night—that I was in the clutches of an evil demon. And he made me do things... like terrible things... to chickens. And then, when I woke up... it was all true. The terrible stuff to chickens, that is. That was all after I drank too much cough syrup with my beers. Did Day-Cart have a Robitussin problem?

Bradley is way off base. For sure, his weird story deserves a moment of reply, but it is best for all involved that a lengthy analysis of Bradley's views on the matter isn't devoted class time. Some views are best left unexamined. Next time, Bradley should read. And lay off the syrup.

In this case, we've offered a kind of moral hazard of iron manning. That is, charity can create dependence. The objective of iron manning, as shown with "Philosophy Student I," was that it should be an instance of showing someone how to approximate an exemplary contribution, for the sake of improving further contributions. But if the student relaxes content with this contribution being a sufficiently good approximation of an exemplary contribution, we've undercut the diachronic good yielded by the iron man. Bradley's paper for the course may be about a philosophically insipid voyage of watching cartoons while high on more cough syrup. And it is worth noting that the same risk obtains with Alfredo and Prof. Zoccolo, too. Iron manning gives greater credit for quality than what the contribution was, and this risks grade inflation with papers and class participation. And it risks the greater problem of inflating the arguer's self-estimation of their insight. Inflating dialectical scores comes with its benefits and risks.

Again, the second issue is that of being able to properly identify whether a speaker and a view is worth the time and effort of iron manning. Because of the inflation problem, it has risks in all cases, it seems. But the moral hazard goes beyond this problem, as responding to improved views takes more time and energy than otherwise. So there is an epistemetric problem with iron manning, very similar to the epistemetric problem for salutary straw manning. If time is short, the cost may be too high, as the improved view may be harder to get out, and so may also be the critiques of it. Participants may take the end of the discussion to be a tie, or that once things got complicated, maybe it's not worth their time. A view iron manned for the sake of complete refutation, then, would survive, but it would only because the discussion was not completed and the view at issue was not refuted. So argumentative backfire (where a view that's not very good comes out looking better than it should, as noted by Cohen 2005) is a risk of iron manning, too.

4.4.4 Defunding the Police

In the wake of George Floyd's death in police custody in the spring of 2020, there was a wave of protests and intense debate about the role of police in American society. Floyd died from being held on the ground, a policeman's knee on his neck for eight minutes and forty-six seconds. The incident was captured on video, and calls for defunding the police were a consequence. The PBS *NewsHour* aired an interview of two representatives of interested groups in the debate.[3] Judy Woodruff, host of the *NewsHour*, interviewed Charlene Carruthers of the Movement of Black Lives, who advocated for defunding the police, and Chuck Wexler, a representative of the Police Executive Research Forum, who advocated for extensive police reform. Woodruff opened by noting that both sides agreed that money and support needed to be directed to community services, and more investments needed to be put into social supports for the needy, drug awareness programs for children, and rehabilitation programs for those with addiction.

Woodruff then asked Charlene Carruthers what the defining issue was for her group. Carruthers pointed out that there were two: first, how much money goes to police departments, and second, how that allocation of resources underfunds struggling communities. If the money that went to policing, surveilling, and intervening in the community were reinvested in improving the community, there would be better outcomes. The money invested in police departments should be redirected, and "band-aid reforms and 'solutions' are not enough." She continued:

> What I would like to see our local governments and police departments commit to is a process in which we move essential safety services from control of police departments to our communities. That means crisis response and that also means how we deal with violence, conflict, and harm. It means that eventually, the police are not the people who intervene or show up in instances of conflict, crisis, or violence. In fact, it's a community-based response that's outside of systems of policing, prisons, and jails.

Woodruff asked Wexler to respond to Carruthers' statement. Wexler's response was that:

> I agree with a lot of what Ms. Carruthers said. This is a joint responsibility with the community—mental health, homelessness, dealing with the opioid epidemic. This can't be done by the police alone. We need the community. I agree with her about investing in the institutions, kids need to have jobs, we need to be working, we need community policing. We need problem-solving. There's not much that I disagree with Ms. Carruthers, except that right now we have a spotlight on the police—that's the time we take a hard look at how they do their job. […] We need community people. They prevent gang violence. They are gang interrupters. […] Now is the time to get into policing, to do problem-solving, to work with the community, and be better at what we do.

Woodruff then asked the question whether the choice is between reforming policing and doing away with policing or *whether the defunding movement really is that different from the reform movement.* Wexler's response was that the two were not very far apart. Carruthers responded:

> Actually, Mr. Wexler and I have a fundamental disagreement. I do not believe we agree on the role of the police. I do not believe we should have a partnership between the community and police in order to deal with conflict and violence in our community. What that point, those decisions, that role, included that money needs to be completely shifted. Not just to social services, because we see most social service agencies as proxies for the police. […] It is not just a few

bad apples. We know the entire tree is rotten. [...] What we are talking about is a big vision. Radically transforming the world we live in. Defunding the police is an essential step.

What is notable about the discussion was that the moderator allowed a speaker to address and correct a mischaracterization. In this case, the issue was whether the defund movement was all too different from the reform movement for policing. If we had heard the interview only up to the conclusion of Wexler's statement of the reform view and his characterization of the defund view, we would likely think that the two movements are not far apart at all. Perhaps that they differ only in emphasis or on details of the redirected funds. But since Woodruff asked Carruthers to reply to the characterization of her view, we were allowed to see the distortion clarified.

Now, both straw and iron man arguments depend on different characterizations of the views on offer for critical reflection, and Wexler's contribution fits the bill for distortion, for sure. But was it a *straw* opponent constructed, or was it an *iron* one? Our answer is that it depends on the audience for the view presented, because the question of better or worse is one that is indexed to the audience for the reconstruction. From the perspective of those for whom the progressive and radical change of defunding the police is a positive one, Wexler's characterization was a straw man, portraying Carruthers as holding another moderate view when more radical change is necessary. However, from the perspective of Wexler and those who hold that there is an essential role for well-regulated policing in a society, the restatement of the defund position as an overstated but reasonable call for reform was an act of interpretive charity. Wexler's recasting of Carruther's view, for this audience, is an iron manning of the position. And this is why it is clear that straw and iron man arguments are not simply dialectical in a sense that there is one speaker representing another's argument (that, for sure, is a core necessary condition) but that the speaker represents that argument *for an onlooking audience*. Straw man arguments regularly are not simple dyadic dialectical relations (speaker to speaker) but are often tri- or polyadic exchanges between speakers and their non speaking audiences. The same can be said of our pedagogical cases earlier, as they can occur in the classroom in front of other students, and so serve as establishing a kind of culture of argument. But they need not be so. In this "Defunding the Police" case, we see something important about the distortions of others' views and arguments—whether they are seen as improvements or damages to the other side's views depends on what the audience for whom the opponent is portrayed thinks about the issues.

There is a particular kind of social epistemic concern that arises in a case such as this. It could be called the problem of *toxic charity* (as noted by Govier 1981; Stevens 2020). It can run like this: in some field or dominant domain of discourse, there is some range of acceptable and appropriate contributions to a critical dialogue. There may be innovations, but these can be made generally internal to the domain, and they take the status of a track record of other well-received contributions to be recognized. So when an external critique is lodged, the problem is how to interpret it—from inside, it seems out of order, silly, or absurd. So to interpret the challenge along these lines is *prima facie* uncharitable. It, as the interpreter reasons, is better to interpret the contribution as one in line with other responsible contributions. And so the radical challenge gets reinterpreted as a call for reform. Interpretive charity, when performed as such, even with the best intentions, becomes a way of talking over the other. And, as it turns out, so can the iron man. The iron man is a form of supererogatory interpretive charity, and just as charity can sometimes be performed as giving people what the giver thinks the people need or ought to need instead of what they actually need, so can interpretive charity be a mismatch of what the best interpretation of a statement from the perspective of the interpreter is and what was said.

Another variety of toxic charity can sometimes be found in misrepresentations intended to repair another's argument. In some cases, such as "Philosophy Student I," these are clearly unproblematic, because the interaction between Professor Zoccolo and Alfredo is in the first place asymmetrical. Zoccolo is an expert; Alfredo is not. Besides, second, Alfredo has also elected to take Zoccolo's class, a certain amount of helping is to be expected and one can reasonably suppose that he has consented to it. Outside of these contexts, we walk among people who are our civic equals. We engage in debates in the public square where an effort to improve or repair someone's argument might be justifiably be interpreted as patronizing or as passive aggressive. "I think what Hapless Arguer is trying to say is that... " or "Perhaps, Clueless Dolt's point might better be put as follows..." This implicates that they're not up to the task of making their own argument, or that they have a weak grasp of their own reasons and so they need your help, even though they didn't ask for it. What might be worse in these circumstances is when the help is not delivered *off stage* but there in front of everyone. Not only can this be embarrassing but it diminishes their ability to make future contributions. Again, in some contexts (the classroom) argument repair is warranted and even laudable, but in others, it poses very serious risks and comes with significant costs. We must tread lightly among our civic equals.

There's going to be disagreement, and there will be cases where arguments fall short, but there is virtue in letting others make their own mistakes and say their piece in their own words.

4.4 Iron Man and Argumentative Virtue

From these cases, the basic form of iron man argumentation can be discerned. First, as a dialectical form, the iron man requires two speakers, our good friends, A and B. A proposes some argument a and/or some position p. But a and p are not defensible. B takes up with A's case with a reconstruction, a* and p*, that given the state of dialectical play are (comparatively more) defensible. Often this strategy is done for the sake of an onlooking audience, C, which may be interested in A's views or the issue of whether that p. So far, again, we can see that there is a dialectical distortion, just as there is with straw manning, but instead of degrading the opponent's argument (as with the straw man), the opponent's case is improved. Hence our term *iron* man.

There are compelling epistemic reasons to regularly iron man one's opposition, as the truth will come out in contexts of maximally responsible and detailed argumentation. Since our epistemic objectives in argument are truth and its understanding, the most intellectually robust opponent is the best, and if one does not encounter but must construct such an opponent, then so be it. Moreover, there are ethical (and political) reasons why iron manning may be appealing.[4] At its core, iron manning is a form of interpreting others' communicative acts with maximal charity. Furthermore, the demands of recognition for underrepresented groups obtain so that their interests can be heard and have effect. Iron manning is in the service of this. Finally, again, there are pedagogical reasons why iron manning may be required, as it sets out exempla for their further intellectual development.

So what, then, could be wrong with iron manning? To start, we hold that there is a fallacy of inclusion for the same reason that there is a fallacy of exclusion.

Let us return to the cases. As we saw with "Philosophy Student II," there are pedagogical reasons why iron manning can be objectionable, as the point of class discussion is for students to improve their own views, not having it done for them. It is here that we begin to see the trouble with some forms of iron man: in taking some poorly articulated views seriously, improving them and submitting them to scrutiny, one makes an investment of time and intellectual energy. The trouble is that there are many investments that are unwise. Consider, further, a feature of discussion after content presentation. There is evidence now that suggests that rude or irrelevant online comments after a posting or story actually distort reading comprehension of the original piece. That is, the more comments that don't get the original point you are exposed to or the more rude comments in the discussion thread, the less likely it is that you will, afterward, correctly recall the details of the posting. This is now being called "The Nasty Effect." Derailed discussion not only is a waste of time but it is miseducation.[5]

Now consider the strategic use of iron manning with the "Eric Cantor" case. The trouble is not with improving the view *per se* but with the way the improvement is deployed. In this case, (a) the iron man is presented as Cantor's view, and (b) thereby it is used as evidence that Stahl is (and liberals generally are) fact-challenged. But this is a distortion not only of Cantor's position but of Stahl's, too. By iron manning Cantor, one straw mans Stahl, his critic. Her criticisms now seem off-target and ill-informed, when they, in fact, were not. This, we've called the problem of dialectical scorekeeping, as the issue of critical dialogue is not simply the propositional result but the process of arriving at the result—the epistemic objectives of argument nevertheless have intermediate goals of keeping track of the contributions others make as there will be other discussions in the future, and we need to keep track of reliable contributors to our inquires. So even if the iron man in the Cantor case may yield higher-quality epistemic outcome in the first instance, it distorts Stahl's (and Cantor's) track record for our estimation of the quality of their work. As we noted before, straw manning and iron manning have diachronic results, too, because they change how we meta-argumentatively assess our discussants as competent arguers.

These two elements of iron manning converge. When one iron mans a poorly presented view, one may encourage those who have posed the view by taking them seriously, and thereby impugn their critics. Again, sometimes this is appropriate, as some views need time and patience for their development and some speakers require maximal charity in interpreting their communicative acts. But sometimes it is inappropriate, as one can be held hostage by these speakers. On blog comment threads and chatboards, there are many who are uninformed and contribute with unhinged criticism. They are out to hijack discussion, to hold forth, to be the center of attention. These are, in internet lingo, *trolls*. Taking the trolls seriously, interpreting them with charity, and responding to them thoughtfully yields only grief. The internet wisdom is right: *one must not feed the trolls*.

Indeed, too often philosophers and argumentation theorists overlook the fact we very often find ourselves having to evaluate just *this* argument from *this* arguer, even if this argument could be stronger, or this arguer could use some help. We have argued here that even charitable alterations of arguments or arguers distort the dialectical landscape and are often unacceptable, for exactly the same reason why straw manning is unacceptable. The only difference is that the straw man excludes arguments that are worth listening to; the iron man includes arguments that are not worth listening to.

Alternately, we have identified a manner of inclusion that is problematic with iron manning. Straw and iron manning arguments have a core of different quality between the statement represented and the representation. On these metaphors, *straw* for lower quality, *iron* for higher. But assessments of quality are audience-indexed, given the dialectical features for this argument type. That is, whether it is a straw or iron man depends on what the audience on the receiving end of the argument representation thinks of the thesis or case presented. And so, as we see with "Defunding the Police," an attempt to iron man an argumentative opponent may change their view so profoundly that it becomes, from their own perspective, a straw man. External critique, when rendered as internal criticism, may be more accessible and even have a kind of standing and uptake, but it erases the voice from the margin in the transformation. And this, even if done with the intention of inclusion, is a form of exclusion. It has both moral problems with a failure of recognition, as dictating to someone the conditions for their regard and their critical voice, but it has additional epistemic problems, as it makes potential defeaters for more global assumptions harder to state and answer. It makes radical doubt about a system more difficult to be posed or be given uptake.

In all, we've identified a few rough criteria for knowing when iron manning is fallacious:

1. When it is clear that the argument to be reconstructed is not likely to be either relevant or successful.
2. When it is clear that the improvement of and response to the argument will take more time than is allotted, and there are other, more clearly salient, issues.
3. When, even if 1 and 2 do not obtain (i.e., when there may be something relevant and there is plenty of surplus time and energy), it is clear that responding to this speaker under these circumstances encourages further badly formed arguments.
4. When the positive reconstruction of the argument (iron man) in question yields misportrayal of the argument's prior critics as attacking a straw man.

5. When the new portrayal is posited on not taking external criticism seriously.

These rough criteria are, in the end, an overlap of (a) issues in cognitive economy (maximizing epistemic efficiency) and (b) issues in maintenance of a properly run dialectical field. We hold 1 and 2 are epistemetric questions, and 3, 4, and 5 are dialectical questions. Hence, the basic thought that sometimes feeding the trolls is (a) a waste of time and energy and (b) it ultimately isn't anything but bad for the way we argue.

4.5 Open-mindedness and Its Limits

We have so far argued that there are reasons to think the wide varieties of straw manning have legitimate or non-fallacious instances. We have also argued that what makes straw manning fallacious also makes overly charitable distortions—what we call iron men—sometimes fallacious as well. We now return to address the challenge leveled by Walton and Macagno, namely that there is no scheme for straw manning. This is indeed true in a narrow sense. The straw man is not a scheme like the *ad hominem* is. Rather, the straw man points to more fundamental argument virtues; in other words, the straw man is about one's approach to argument as a whole, rather than one's employment of single argument schemes. And the same, by parity of reason, goes for the iron man.

There is no mistaking the fact that the straw man in its core forms is a kind of mistake. It's one of the unhappy families in logic, on our *Anna Karenina* image. They're all different. But getting arguments right—well, that, in a way, is the happy family. And they're all alike. Well, we've argued here that this is not so. But nonetheless, to improve a view is more like a virtue than not. And so we think it is worth looking at iron manning through the lens of virtue theory, where it stands poised between excess and deficiency. The appropriate virtue, we think, is open-mindedness.

It is a simple truth that open-mindedness is valuable, that it is an intellectual virtue. One should be ready to consider new and different possibilities, one should be open to being surprised, and one should acknowledge that one is fallible and so not dogmatically reject views inconsistent with one's own as obviously false. Open-mindedness is not only an intellectual virtue but a particular kind of intellectual virtue, one bearing on dialogue. To be open-minded means that one's openness is to other ideas, lines of thought, and objections. Consequently, it is a dialectical virtue.[6] One may, for sure, display it monologically (perhaps with prolepsis), but to do so is dependent on there being antecedent alternative views to which one must be open and responsive. In that case, it must be in the first instance dialogical.

Open-mindedness has a strong resonance with both the variety of straw man and iron man forms. Virtuous straw manning, say with proposing a bad version of a dialogue partner's views in order to allow them to clarify the commitment, displays a kind of intellectual curiosity that dovetails with open-mindedness. It is as if to say *the view surely isn't this (terrible view), so show me what it is*. The same can go for the other forms of straw manning. One may criticize a widely held but poorly informed view held by opponents so as to have better focus later on for better informed views, and so virtuously weak man the opposition. And one may invent a ridiculous version of opposition in order to identify errors that are easy and tempting in the dialogue, and so virtuously hollow man.

We noted earlier that there are clear reasons why one should consider the act of iron manning opponents to be virtuous. For the sake of stable critical dialogue conclusions, the best versions of views should be considered. If the view can be improved after the discussion, the conclusion will be unstable, and so critical discussion must be opened anew. For the sake of the time and energy involved, it is best to get as far as one can dialectically in one's exchanges, so one's defaults should be on best versions of views at issue. Iron manning, then, is a useful article in the virtuous arguer's toolkit. Sometimes, the opposition's views are not as developed as they otherwise could be. Perhaps it is because the view is new, and so still fuzzy in the minds of its adherents.[7] Or some of the evidence for the view is tangled and complicated, so there is no easy way to see how the case for the view should go. Or the view is held by a minority who, because of social pressures or outright prejudices, are not heard or have not had the opportunity to make the view clearer. There are not only purely epistemic reasons to default to iron manning but also ethical and political reasons. One must, as a matter of justice and charity, try to hear and understand marginalized voices. Social privilege can carry with it the illusion of epistemic privilege. Open-mindedness, as a state of intellectual character, is a virtue that reflects both the epistemic and moral importance of being willing to hear out and charitably interpret alternative views. Iron manning is how that virtue is manifested.

However, we have seen that iron manning has three pitfalls. The first is the overall dialectical change that comes with iron manning—other critical responses to a view under scrutiny, once the view is strengthened (we might say *cum ferro*[8]), are made to appear as representational straw man criticisms. It may be a favor to the criticized view, but it is unfair to its other critics. It seems appropriate, for dialectical clarity, to make coordinating moves, to highlight the older and weaker view and its appropriate critics and then to present the view *cum ferro* and explain how the improved version is no longer an appropriate target for the criticisms.

The second pitfall of iron manning is what we'd termed the epistemetrics of dialogue—some views, even when improved, are not worth the time and energy for full hearings. The challenge for open-mindedness is that it seems that the charitable side of charitable criticism has outweighed the critical side. Must the physicists stop all they are doing to give a full hearing to someone who thinks that they've made a new cold fusion reactor or created a perpetual motion machine with kitchenware and spare lawnmower parts? Does a dietitian need to give the advocate of the Corn Chip Diet hours to present their data? Should we give every ghost sighting and alien abduction story full credence?

Related to the epistemic reasons is a further consideration, which is the broader moral hazard risk with iron manning. As we'd noted, it can create an illusion of competence, as shown by the phenomenon of grade inflation attending iron manned essays from underprepared students. For sure, we do not want to crush the spirit of those contributing to discussions, but often others are in need of some negative feedback on their performances. Else they will harbor the thought that because they never were called out for their bullshit, they were not full of it.

The third worry about iron manning is the problem of toxic charity, where the change in view may be an improvement from the perspective of the one iron manning and perhaps even their audience, but it erases the salient information crucial to the speaker's message. So, as we showed with "Defunding the Police," a more radical view, iron manned by a moderate for a moderate audience, becomes a moderate view. And so in doing a favor for a speaker, iron manning nevertheless does little favor.

What is required in these cases is the appropriate deployment of three categories[9] for claims: (a) those claims with greater evidential support than their competitors, (b) claims with less evidential support than their competitors but that have some plausibility and other considerations in their favor, and (c) claims without evidential support and for which it is unlikely support will surface in further inquiry. Let us call these *well-supported*, *debatable*, and *indefensible* views, respectively. Being open-minded means one not only deploys these categories but deploys them accurately and responsibly. And it is with the last class, the unsupported and indefensible views that open-mindedness has its limits for time and patience. As William Hare puts it, "open-mindedness does not mean pursuing every will-o'-the-wisp" (2003: 82).

Given that we are using the concept of *virtue*, it may be useful to wax Aristotelian and invoke the notion of the *mean between extremes* for open-mindedness as a virtue. Just as *courage* is the mean between cowardice and rashness and *generosity* is the mean between miserliness and profligacy, open-mindedness can be understood as a mean between two coordinate vices. The vice of deficiency in this case is *dogmatism*. One is insufficiently charitable to and unwilling to take seriously alternative views. The vice of dogmatism, then, not only rejects the third class of views, the indefensible, but hastily rejects members of the second class of debatable views. And notice further that the vice of dogmatism is what is behind the problem of toxic charity—namely that legitimately framed objections or critical points can be posed only from within one's preferred vocabulary. Reinterpreting challenges from the margin in the voice of the dominant discourse is a broader conceptual or cultural dogmatism about epistemic pathways of those brought up within that discourse.

We have no easy term for the vice of excess in the case of dialectical charity. In a way, it is the dialectical correlate of the doxastic vice of excess of belief, that of *credulity* (as opposed to the vice of deficiency some might call *skepticism*). In the same way that the credulous person uncritically believes many things unworthy of assent, those with the vice of excess of dialectical charity uncritically devote time, energy, and resources on undeserving and indefensible issues. Every kook and crank gets a full hearing, and perhaps receives such a charitable reading that their views are reconstructed into better, comparatively responsible *cum ferro* versions. This, again, has costs, as many do not learn to themselves improve with this act of charity, but blithely believe that they are doing excellent work, given how seriously they are being taken.

Given this rough Aristotelian triadic version of the virtue of open-mindedness, another insight from Aristotle's model may be imported. Aristotle had observed that the mean often requires, at least from the perspective of the agent, overshooting the mark to compensate for inherent proclivities toward one kind of vice. So, with generosity, we generally should try to overshoot on the side of profligacy, since we are naturally disposed toward deficiency. Alternately, we should overshoot toward rashness with courage, since we are usually disposed toward excess of fear in cases of genuine danger. This *doctrine of compensation* for the virtues is captured by Aristotle's image:

> We must draw ourselves away in the opposite direction, for by pulling away from error we shall reach the middle, as carpenters do when they straighten warped timber.
>
> (NE 1109b.5)

In dialectical matters, our own crooked timber curves toward dogmatism, a deficiency of charity. As a consequence, the default on iron manning is dialectically virtuous in the first instance of improving dialogue quality, as we noted in the previous section, but in the second instance of compensating for our natural deficiencies of dismissing views we do not hold. There is, of course, a paradox internal to the advice that one try to overshoot what one takes as the mean. It requires that we make a judgment about the mean definitive of virtue, and then try *not to* hit it. This seems, on its face, an instance of intending to be, on Aristotelian terms, vicious. Of course, it is not what it comes to, in the end, but it's worth pausing to clarify our view about correcting for our own crooked timber and why argument is a key place for that. In essence, such correction requires that we see ourselves from the third-person perspective on many matters, as the problem arises from our uncritical and unreflective inclinations to believe what we just take to be true from the first-person perspective. We live our lives from the inside, both with what we value and what we believe. And we believe what we believe because we take those things to be true—what else would it be something to believe? But we are fallible creatures, and we discover that fallibility when we are presented with a failure, a surprise result, manifest falsity with what we'd held, a misfire. And so as we retrace our steps, we can see what kind of evidence or inclinations led us astray. But doing this reverse-engineering of a mistake requires that we toggle between the perspective of the person who was thinking it through for the first time and in the first person and the one who knows where the path leads and can tell that story in the third person. We can make notes about what might lead people who think about these matters astray, and we can identify things that lead us in particular off the track, too. And we do so in order to correct for this in the future. Just as it is perfectly rational, if not slightly contradictory, to commit the paradox of the preface, standing by all the statements in the book, but also holding there must be errors, we know that we hold false beliefs, but we cannot identify them individually. What the virtue of open-mindedness allows, then, is an openness to allowing others to help with the correction of that crooked timber. Their dialectical pushback bends us in argument, when appropriate, to what is the true mean. Proper use of the iron man is the argumentative touchstone for open-mindedness.

But in what way is the excess identified and corrected? The old trouble returns for the open-minded, especially in light of the doctrine of compensation: How do we keep open-mindedness from being empty-headed, uncritical credulity?

The answer is that no virtue stands alone. Open-mindedness must also be coordinated with seriousness, clarity, and wisdom. Just as courage is ultimately not a good when it is exercised without justice and generosity is not a full virtue without judgment as to how and to whom to be generous, open-mindedness is no virtue without other deliberative virtues. When time is short and intellectual resources thin, more conservatively framed discussions are in order. They are the safer bets. But when there is time enough and resources, liberality with discussion is appropriate. It is, perhaps, not a surprise that virtues are themselves virtues only in the company of other virtues, moral and intellectual. So it is with open-mindedness and its hallmark, the iron man.

Though open-mindedness is a default virtue, it is because we have the thought that our intellectual environment, our sources of information and those with whom we interact and exchange information, is generally reliable, cooperative, and amenable to getting at the truths we care about. That is, open-mindedness is a virtue when the situation is amenable to following evidence and exchanging reasons honestly. But in environments where these defaults are not the case, say in environments polluted with misinformation and others bent on manipulating us, open-mindedness is an unacceptable form of intellectual vulnerability (see Battaly 2018; Kidd 2019). And, for sure, iron manning others in that kind of environment would be an error.

4.6 Taking Stock

We have argued in this chapter that the dialectical phenomenon known as straw manning is much more varied than many accounts suggest. In the first place, straw manning involves more than simple distortion. It also includes forms of selection (weak manning) and invention (hollow manning). Second, not all instances of straw manning are fallacious. Third, and somewhat ironically, charitable variations on an argument suffer from the same failings as fallacious straw men, though their mistake lies in the inclusion of arguments deserving of full criticism and rejection, and this we've called the iron man. Finally, we have shown that these argument forms and their virtuous and vicious forms allow a clearer picture of the virtue of open-mindedness and its limits.

To this point, our theoretical approach has been to describe a complex phenomenon and then evaluate the instances. So, as fallacy theory, we've shown two things—that the fallacy is common enough to require its own category (and sub categories) and that there is a normative account of what goes wrong in the case of these particular fallacies. Further, we've shown that there can be cases wherein these argument types can be appropriately used. The key is that with fallacy theory, that's only half the job. Observing that there's a common error is, at this stage, only the beginning. The next part is to give an explanation of how the error happens and how to correct it—that is, how the illusion of argument quality is created and how to dispel it. At this point, then, we must make a transition, from descriptive and normative, to the explanatory and prescriptive. So, in the next chapters, our objective will be to address the puzzles regarding the straw man fallacy, the *effectiveness, meta-argumentation,* and *dialecticality* puzzles. With them, we think a clear explanatory and prescriptive story emerges.

5

The Puzzle of Effectiveness

Up to this point, we have shown a number of things. They include that the straw man admits of a variety of forms: the standard straw man, the weak man, and the hollow man; that these forms, though generally fallacious, have argumentatively salutary, non-fallacious, instances; that just as straw manning is a fallacious misrepresentation of a dialectical opponent as worse than they actually are, there is a correlate fallacy of representing one's opponents (and allies) as better than they actually are, the iron man; that the iron man, too, admits of fallacious and non-fallacious instances. And further, we've shown that audience selection plays a significant role in how straw and iron manning works, as both argument quality and distortion detection depend on the audience's comportment toward the instances and the opponents represented. And we've argued that, because there are argumentatively salutary straw, weak, and hollow man arguments, the fallaciousness of the straw man does not reside in the misrepresentation or in the refutation of the presented straw views but in the meta-argumentative function of these arguments prematurely closing the inquiry. Finally, we've brought out a feature of the straw man and iron man that highlights background norms and attitudes in cultures of argument and microclimates of critical exchange—namely, that the non-fallacious instances of these forms are made possible in cases of trust, and heightened distrust and enmity intensify fallacious versions. We think that the straw man and its varieties are theoretically fecund, and identifying this variety is an occasion for some deeper reflection on the norms of dialogue that the straw man highlights, particularly in its fallacious forms. In the service of this broader program, we are set to address the orienting puzzles of the straw man, what we've called the puzzle of effectiveness, the meta-argumentative puzzle, and the dialecticality puzzle. Our plan in this chapter is to address the hardest of the puzzles, that of effectiveness, and then turn in the following chapter to the puzzles of meta-argumentation and dialecticality.

5.1 Clarifying the Puzzle of Effectiveness

We have shown that the straw man fallacy admits of a wide variety of forms, ranging from what we've called the weak man, to the burning man (later in Chapter 6), and even to the iron man. What makes all these different forms instances of the same general kind is the dialectical core of the fallacy—the misrepresentation of the argumentative state of play between contesting sides. In most cases, one side is represented as argumentatively worse off than they actually are (though, in cases of iron manning, one improves an interlocutor's case). Again, it is this dialectical core of qualitative misrepresentation of views and reasons that makes straw man fallacies as a class distinct from, say, fallacies of relevance like *ad hominem* abusive or arguments from pity. In fact, what's interesting about straw man arguments is that they are, really, *arguments about arguments*. In other words, when we argue, we can commit particular kinds of fallacies, but unique kinds of fallacies occur when we reason about how we reason. They are fallacies rooted in and made possible by our metacognition. By our lights, all three puzzles about straw man arguments, effectiveness, meta-argumentation, and dialecticality, are all of a piece, but are distinct in terms of stating the simple fact that the core of straw manning requires representing the thoughts of others. The place to start, we think, is with how it is possible for a misrepresentation of the thought of an other can be effective as an argumentative strategy. We'll then look at its deeper implications with regard to dialogue, argument, and reasoning about reasoning.

A long-standing, and perhaps obvious, problem with straw man arguments is that when they are presented to the target of the straw manning, they typically are ineffective. We generally can tell when an interlocutor has misrepresented our view. The straw man directed at you at best can function as a signal that your argument is hard to understand or that your interlocutor is dense, but when a straw man of your view is presented *to you* it is unlikely to change your mind about how things stand. One wonders, then, how straw man arguments function. This is the effectiveness puzzle—it seems that straw man arguments, when we present them, should never work. But fallacies are supposed to be widespread and psychologically tempting. So what gives? Our first answer is that straw man arguments of one type do their rhetorical work not on the target depicted as made of straw but rather on an audience of argumentative onlookers, often selected specifically for the argument by the straw manner. There are other cases we will consider that require we help ourselves to the social epistemic concepts of *gaslighting* and *silencing* for when straw men arguments are addressed to their targets in the second person (as *you*). To start, let's make a distinction between straw man arguments given in the *second* and *third* person that tracks this difference.

Straw man arguments in the *third* person require three interlocutors, or participants. There is the *speaker*, the *target*, and the *audience* for the straw man argument. And so the speaker represents the target's views for the audience. The grammatical form of the straw man is that the speaker must say something like, "*They* say..." So, *third* person. The straw man in the *second* person requires only two interlocutors, the *speaker* and the *target* for the straw man representation. So the grammatical form of the straw man in this case must be that the speaker says, "*You* say..." So, *second* person. This distinction is important, as we think that the solution to the effectiveness puzzle for straw man arguments will be different, given the different forms, who is being addressed and who is being represented. The problem is significantly easier in the third person, but it's worth a significant amount of time to theorize properly, because it speaks to our other point about salutary background cultures and virtuous argumentative microclimates. The problem is harder to solve for the second person form of the straw man, and it is, again, here that we will need to use the notions of purely dialectical wins to yield *silencing* an opponent and that of *gaslighting* as our primary explanatory tools. We will begin with straw manning in the third person. Consider the following straw man in the third person.

5.1.1 Biden Defunds

In the run-up to the 2020 presidential election, US President Donald Trump consistently portrayed his Democratic rival, Joe Biden, as having supported a program of *defunding the police* in the aftermath of police killing Breonna Taylor and George Floyd, among many others. In fact, Joe Biden had not supported these measures, and had instead proposed plans for police reform. No one familiar with Biden's position on the matter will be moved by Trump's representation of his views in the following tweet.

> Corrupt Joe Biden wants to defund our police. He may use different words, but when you look at his pact with Crazy Bernie, and other things, that's what he wants to do. It would destroy America![1]

Again, this highlights the curious feature of straw man arguments—the target of the straw man argument is bound to see the purported criticism as misplaced, even fabricated. However, although straw man arguments are *posed* as criticisms directed toward one's interlocutor, they are actually aimed at third parties, an audience that is either antecedently disposed to accept the straw man version of the interlocutor or else sufficiently unfamiliar with the matter in dispute to fall for the misrepresentation. The key is that the straw man argument requires that its audience either not know or not care that there is an ill fit between the representation of the opponent's view and what the opponent's view actually is. Relying on one side of the dispute to supply an account of what the other side thinks is a recipe for all-too-convenient distortions.

Next notice a further thing about this particular instance of straw man—it depends on its audience *vehemently rejecting* the plan of defunding the police. For a large majority of Republicans, the proposal to defund the police is a bridge too far, effectively a proposal of anarchism. (And it is worth noting that Biden's campaign has thrown the accusation back at Trump, with this same audience's views in mind.)[2] But it should be clear that Biden's own moderate position in the debate is one of consternation among progressives in the Democratic Party. In fact, were Biden to propose a defunding program, it would be enthusiastically embraced by a good many progressive voters.

The lesson, then, is that straw man arguments in the third person work not by tricking an interlocutor into accepting a distortion of their own view but rather by addressing a selected audience of onlookers. Moreover, straw man arguments select their preferred audiences in two ways. First, the audience must be listeners who will not register the difference between the misrepresentation of the opposition presented in the debate and the actual views of the opposition. Second, the argument must select the audience that will view the constructed view and argument as *worse*, instead of *better*, than the actual view or argument of the target of straw manning. When Trump straw mans Biden on police reform, with the defunding accusation, he does so for an audience who must see this as a bad plan rather than a good one; moreover, the audience must be either inattentive to Biden's actual proposals or antecedently believe that Biden does in fact back something along these lines.

Note a further upshot. When straw man arguments succeed, they convince an audience that the opposition is so benighted, dishonest, untrustworthy, or unreasonable that one may simply disregard what they say. A visit to the comments section on Trump's tweet from those who identify as his supporters bears this out, as does a visit to the opinion section at *Breitbart*, a publication strongly behind Trump's re-election, after the tweet. This has two related effects. First, attempts by the straw manned interlocutor to correct the distortion are bound to fail, simply because those who have fallen for the straw man either are no longer listening or are no longer disposed to trust what the interlocutor says. Second, once the target audience has accepted the straw man depiction of the opposition, one has free rein to construct further and even more outrageous distortions. Straw men breed quickly (and we will address this phenomenon of "bevies" of straw men as *burning man* fallacies later in Chapter 6). Often there's little that can be done to defend oneself except to respond in kind, by manufacturing more straw men or not to feed the trolls. Accordingly, even those who want to play fair and address actual issues in dispute are thus dragged into the straw manning game or just begging off arguing altogether.

Straw man arguments seem ubiquitous, and though it is clearly bad news overall, there is a slice of good news in them. In criticizing others' reasoning, we show a care for good reasoning. Straw man arguments would be impotent were we not to value cogent arguments and share tools for criticizing arguments that fail those standards. The bad news, of course, is that we all too regularly fail these norms, both in reasoning about things and also in reasoning about how others reason.

A long-standing rule of thumb to avoid falling for straw men is simple: don't instantaneously accept your own side's depiction of the other side's views. Go to the source instead. However, in the political context, with the pressure of the perpetual campaign, candidates are incentivized to fix nearly exclusively on their opponents' flaws, and sometimes, they must invent them. And given that time is short, the window for correction is quite small.

The lesson, as we see it, then, is that we have arranged some useful explanatory tools to address the effectiveness puzzle for straw man arguments in the third person, as they depend on appealing to an audience that that is unfamiliar or unsympathetic with the target for the straw man. And it is a feature of our epistemic dependence that we often must rely on others for information about the views of those they criticize—often the targets are not present, and perhaps they are unwelcome participants. So our epistemic dependence on others opens us to these dialectical manipulations. Consider, next, the following instance.

5.2.2 Fox Fascists

On February 2, 2017, Fox News's Twitter feed had a post with a picture of graffiti on a wall. In large, red, capital letters, the graffiti read, "NO FASCIST USA." The tag on the picture read, "Anti-Trump graffiti spree a sign of emerging alt-left." The reaction to the Fox News tweet from the left was that Trump was not mentioned in the graffiti but only fascism was, so why did the folks at Fox News think that this was anti-*Trump* graffiti? In re-tweeting the Fox tweet and commenting on it, journalist and political commentator Benjamin Dixon's read was: "So we agree that Trump is a fascist," and "they told on themselves." Comments on Dixon's tweet echoed the sentiment—that it was a "well-expressed Freudian slip," that the Right "self-owns every other day," and "if the jackboot fits...."[3]

The reading that Dixon and the commentators have of the line from the Fox News Twitter feed was that in reading anti-fascist graffiti as thereby anti-Trump, Fox News conceded that Trump is a fascist. But the problem is that news reporters (as most speakers of languages) are capable of *indirect discourse*, not just using their own interpretive lenses to make sense of what others say but using the interpretive lenses of others to make sense of what they say. So, for example, if Vihaan knows that Swathi thinks that socialized medicine or regulations on food preparation promoting healthy options is a form of the "Nanny State" paternalizing its citizens, he can make sense of Swathi's statement, "The Nanny State is now making salty fries illegal," without conceding that such a policy is a form of a "Nanny State." He knows what *she* means by that term, and so can interpret *her* claims without himself conceding the identity. That's how indirect discourse works. So when Vihaan says, "No, Swathi, the Nanny State isn't coming to take your salt shaker away," Swathi would not properly interpret his sentence if she replies, "AH! So you concede that there *is* a Nanny State!"

Since the reporters at Fox News are aware that the "alt-left" considers itself anti-fascist and opposes Trump, and further equates Trumpism with fascism, they have good grounds for interpreting statements by those identified as the "alt-left" about resisting what they term fascism with resisting Trumpism. Trump is the occasion for their invocations of fascism. So, it is not because the Fox News reporters *themselves* must take Trump to be a fascist that they are able to make this connection, it is because they know that *the speakers they are interpreting* make this connection that they interpret this graffiti as they do. Again, that we are capable of indirect discourse is how we are able to do that, and to lose track of that makes it so we make the error above.

This would be a form of straw manning that amounts to, as Walton and Macagno phrase it, "wrenching from context" (2010). In this case, the context is the knowledge of how a group uses the term "fascist" in connection to criticizing the Trump administration. But the key to the interpretive error by Dixon and his readership is not simply in mistaking indirect discourse for direct but in taking this particular instance of indirect discourse as uniquely telling. Again, Dixon says that Fox News "told on themselves," and one commentator calls it a

"Freudian slip." The important part here is not simply that there is an attributed error but that this error is revelatory of a wide set of dark truths untold. Of course, for all of this to be plausible, the indirect discourse must be taken as direct, and the error here must be taken to be revelatory instead of a simple error or misspeaking but of a deep, dark secret; there has to be a background of antipathy and mistrust between parties. And it's important to see that Dixon's re-tweet was not for the sake of the Fox News broadcasters. In fact, the key is that Dixon uses the *third person* when representing their view. That is, he represents the statement in the Fox News tweet not for *them* but for an onlooking audience, namely *his* twitter followers. These readers, he knows, are reliably left-leaning and are generally critical of Fox News's reporting. And they themselves strongly suspect or take themselves to know that Trump is a fascist. And they strongly suspect that the folks at Fox News know it, too, but just don't care, are OK with it, or even endorse it. So they are keen to see more confirming evidence to that effect. The convenient, if not distorted, read of the Fox News tweet was a welcome occasion for that bit of hermeneutics of antipathy.

In this case, we think this is a standard form of representational straw man fallacy—the interpretation that the Fox News tweet conceded that Trump is a fascist with its read of opposition to fascism as opposition to Trump is an instance of a misrepresentation of an opponent's standpoint with the result that it makes the standpoint manifestly easier to criticize. And it is important that the representation and the criticism are not presented to the represented and criticized party—the representation is in the third person, not the second person. All of the forms take "they" when attributing the equation of Trump with fascism, not "you." The reason why is that this instance of straw manning is not presented to the criticized party but to an audience that is keen to see the criticized party dealt a deadly critical blow. In this case, it's in seeing them make a statement, interpret the statement in a way that makes it a slip, and then infer that this slip reveals a deep-seated but unspoken view held by folks in the Fox newsroom.

Let's pause for a moment to develop this notion of *hermeneutics of antipathy*. The strategy is to inculcate an ethic of expecting the worst from a particular target group, and the straw man on offer both takes advantage of and contributes evidence to this interpretive stance. It is a familiar-enough phenomenon for many of us, at least on an individual level. We may meet someone who just rubs us the wrong way, and when they say anything, it just sounds wrong. You may dislike the person so much that you may just *want it to be wrong*. That's the hermeneutics of antipathy. And so, your target may say something uncontroversial like, "The sky is blue," and you'll respond:

Well, actually, sometimes, it's *red*. That's called *sunrise* and *sunset*. Sometimes, it's *grey*, that's called *overcast*. And sometimes, it's *black*, and that's called *night*. Nice work overgeneralizing from clear sunny days, jerk.

And there we are. Hermeneutics of antipathy has it so that you'll work overtime to find a problem with a person's statement, even the anodyne ones (what are they hiding?). The hermeneutics of antipathy are the result of antecedent negative attitudes influencing how we interpret, scrutinize, and reply to statements from others. We interpret them uncharitably, we hold them to a high degree of scrutiny, and we reply to them with the objective of not just undercutting or rebutting the view but impugning the intellectual and moral character of the target. It's the result of the oppositional-adversarial core of argumentative exchange. If you're arguing with someone, it's likely because you disagree, and argument makes you and your views vulnerable to their falsities and vices. So this hermeneutic device is an insulating one, as it renders whatever challenges they may pose to you and your views inert. These opponents merely make noise or only speak falsities and nonsense. And this set of attitudes is easily shared and inculcated within a group with ideas and ethical views that are the salient features of the group's identity—it's very important that those who have challenges to those core commitments are viewed as threats only in one sense (perhaps as mere raw power or untoward temptation) but not threats in terms of truth, justice, or rationality.

And so, we have grounds now for interpreting the strategy of invoking, in the American context, *Libtards*, which runs "liberal" together with a slur for the mentally challenged. And *Christofascists*, which identifies a religious political identity with an orientation toward totalitarianism. These instances do not simply describe the opposition, but do so in a way that makes them impossible to take seriously. And they are expressions of the contempt and enmity that the sides hold for each other. In the case of "Fox Fascists," the orienting feature of the interpretive comportment is to start with the assumption that we are reading statements from the morally or intellectually bankrupt (or both). When one starts with this attitude, one will certainly find what one is looking for quite easily. And there will be a ready audience for its consumption.

As we'd noted earlier, a key to straw manning in the third person is to serve up interpretations of the target to the audience that confirms their antecedently negative views of them (or to use the hermeneutics of antipathy to inculcate that comportment in a neutral audience). We've shown this point with how indirect discourse can be interpreted, but this can also be done in terms of how the target's track record of choices is described. Consider the following way you can straw man someone. Pick out a bad decision they made, and then say they *chose* that bad part of the decision. See the following for example.

5.2.3 Spider Cabin

Say my wife and I are trying to decide where to vacation. She wants to go to a cabin in the woods—something rustic and woodsy. But we get there, and the cabin's filled with spiders and there's a hissing opossum in the fireplace. Angrily, I say: *We could have gone to Chicago, but you preferred a cabin filled with terrible varmints and vermin!* Yes, that's the choice she made, but not what she chose *as she chose it*. What she chose was *rustic vacation*, maybe with a hike by a lake and some nice autumn leaves, but what that choice yielded was creepy-crawly arachnids and a hissing and hideous marsupial. The lesson: our desires are propositional attitudes, and those attitudes represent what we desire or choose under a specific description. Again, she chose rustic cabin and it happened to have spiders and a hissy possum. She didn't prefer the varmints. She just chose something that turned out that had them. That's not choosing spiders and opossums. So it's a straw man—you're misrepresenting the intentions of your interlocutor by describing them under the description of their worst consequences. It's certainly not a way to make it so my wife will want to plan another vacation. Or if she does, maybe she'll leave me at home. So now that the point about choice under a description and straw manning is clear, let's turn to another example.

5.2.4 Obama's Middle East

George Neumayr, in September of 2013 at *American Spectator*, portrays the Obama administration's turn on foreign policy, particularly that in the Middle East and Syria in particular, in negative lights. His view is not just that they make bad decisions but that they *choose* terrible things.

> Ho Chi Minh once said that he won the Vietnam War not in the jungles of Asia but on the streets of America. Islamic terrorists could make a similar claim: from Libya to Egypt to Syria, they rose to power not in spite of American leaders but because of them. Obama and McCain preferred Morsi to Mubarak, the assassins of Christopher Stevens to Gaddafi, and now the enforcers of sharia to Assad.[4]

The final point about Syria is a familiar one, as at the time, there were no clear good foreign policy options.[5] The point is that there would be an unintended consequence of destabilizing Assad—the opposition's not a bunch of liberal-minded democrats but radical Islamists. At the time, the ISIL caliphate was on the rise. And it's not that with the Arab Spring, the Obama administration *chose* to support a member of the Muslim Brotherhood to lead Egypt, or that there would be a terrorist attack on a consulate in Libya. Those were the consequences of the choices, but, again, choices are under descriptions, and not all consequences are the descriptions of what was chosen.

Another way to look at this would involve alleging one is committed to the foreseeable logical consequences of one's propositional attitudes. So one can attack the commitments of government shrinkers as "granny starvers," as that's what that view entails. But one should weaken the notion of "foreseeable." Consider the "Spider Cabin" choice, again. Spiders are *foreseeable* in one sense, in that it is possible that there are spiders in a woodsy cabin. But they weren't in another sense, in that one has positive evidence that there are spiders and it's a matter of considering that evidence or not. For sure, afterward, we've got hindsight bias and attribute the latter kind of "foreseeable" to the one we accuse, but it's more likely that we've got a case of a decision with consequences in the weaker sense of "foreseeable." So the description appropriate for a belief or choice must, for accuracy's sake, be of the belief or choice under the description of the one making the choice or under the description of what was foreseeable from the perspective of the person making the choice. The problem, again, is that "foreseeable" is vague, because "able" is a gradable concept. There is one sense that, on "Spider Cabin," the vermin problem is foreseeable, in that rustic cabins *can* have pests. But unless there is reason to suspect this cabin *will* (say, given the age of the cabin, reviews on Yelp!, or maybe that the cabin is named "Spider Cabin"), then one is not negligent for not expecting them. That's the *strong* sense of foreseeable, which is that one doesn't have available evidence to rule out the possibility, and given that there is evidence for that possibility (given that it is actual), it was in this sense foreseeable. Then there is the weaker sense of foreseeable, which is that there is available evidence to the decider at the time of the decision that this possibility is likely. In the former, *strong*, sense, it would be a straw man of a decider's choice to say the decider *chose* that path. So to say that my wife *chose* to go on a vacation with spiders and possums would be inappropriate, because those were foreseeable consequences only in the *strong* sense. But, for example, to say that someone who wants to shrink government and governmental support for the social safety net is a "granny starver" is a little more on the nose, because those are foreseeable consequences in the *weak* sense. Our small government advocate has plenty of evidence for what the consequence of this decision is—he just doesn't care. But, again, we don't think that's the case about the vacation planner for the spider cabin.

Now, returning to "Obama's Middle East," the question was whether the choices under description are results that are only strongly foreseeable or weakly so. It is certain that in the Libya case, it can be read only as strongly foreseeable, as it was not clear that Gaddafi's overthrow would result in an attack on the US consulate, and the same, arguably, goes for Morsi in Egypt, as the evidence initially showed a democratic result for the Arab Spring. And as to the matter of Syria and the developing ISIL Caliphate, saying Obama *chose* that is far too strong, as it was a better point to say that the Obama administration saw it as a better result to put the Assad regime and ISIL at odds (and so choosing *both*, in order that they may cancel each other—again, not a good choice, but maybe the best of a bad lot).

The details of the decisions, however, were not particularly important to Neumayr, as they were, in a significant way, beside the point. The issue was about making sure that the reliably right-leaning audience for his article at the *American Spectator* identified the results as bad and were keen to attribute it as Obama's *intention* to do the bad thing. Tracing the track record in terms of not just bad *results* but bad *choices* (as, again, the bad results of the choices were the targets of the choices) is the key to the straw man on offer. It is an exercise of the hermeneutics of antipathy. Anyone familiar with any of the details of the Obama administration's decisions over that time would object that things were either more complex or muddier than this picture paints, and it would be certainly inappropriate to say that the administration *preferred* that an ambassador was assassinated. But Neumayer's audience, again given the workings of the hermeneutics of antipathy, has no interest in these details or, if they do, they think they are more of a smokescreen for the Obama administration's vices. Again, one signals to one's preferred audiences when straw men are constructed in the third person.

For sure, this audience-indexing effect can still be in place if the *performance* of the straw man argument is in the form of second person, at least grammatically. So, imagine that the grammatical form of the argument is in the second person, but the straw man is not actually presented to or for the target addressed. Instead, even though it's said-to the target, there's actually addressed-to a preferred audience. Giving the argument to the target of the straw man is more argumentative theater than actual argument. Presidential debates run in such a fashion. Consider the following from the recent 2020 US presidential election.

5.2.5 Judges

In his first presidential debate with Joe Biden in September of 2020, President Donald Trump made the following charge against Biden's record as vice president in the previous Obama administration. Trump made the comparative case that he, himself, did the job of president, filling judicial appointments, whereas the Obama administration and by extension, Biden, had left many positions open upon leaving their post.

> The job that we've done—And I'll tell you so some people say maybe the most important, by the end of the first term, I'll have approximately 300 federal judges, Court of Appeals judges 300. And hopefully, three great Supreme Court judges, justices. That is a record, the likes of which very few people—And you know one of the reasons I have so many judges? Because President Obama and him (Biden) left me 128 judges to fill. When you leave office, you don't leave any judges. That's like, you just don't do that. They left 128 openings. And if I were a member of his party, because they have a little different philosophy, I'd say "if you left us 128 openings, you can't be a good president. You can't be a good Vice President."[6]

The problem, of course, is that Obama and Biden were not able to fill any of the judicial openings for the last six years of the Obama administration, because these positions require Senate approval, and the Republican Senate Majority Leader, Mitch McConnell, refused to allow any appointment hearings for any Obama judicial nominations. And so hundreds of positions and one Supreme Court seat sat empty until Trump took office. And Trump then had a cooperative Senate for those empty appointments. So it was not because Biden and Obama failed to do what they were supposed to do but because the Senate refused to do what they were supposed to do that created the unique deficit. And so, when Trump says of Biden and his role in the Obama administration that they "left us 128 openings," that incorrectly states what the situation actually was—the Senate and Mitch McConnell gridlocked the nomination process to the point where there weren't any nominations possible. Of course, this is not exactly constitutional, as the issue of separation of powers is that the legislative body can approve or reject the executive's appointments, but this is supposed to track the institution of government's division of powers, not that of party affiliation between the branches of government. The current reading of it all now is that the Republicans' rule makes it so that the president can make judicial appointments only if the president's party also has a majority in the Senate. Otherwise, there are no appointments. This, of course, is not to be laid at the feet of the outgoing Obama administration, who were targeted by this abuse, but rather the Republican Party, who conspired to have it all happen. That Trump got to fill these positions is as little a testament to what an excellent president he was as the fact that these positions were left open is testament to what a bad vice president Biden was. In both cases, again, it's little. Biden himself surely knows this and any supporter of Biden would know this and be sympathetic to the reply. But Trump, though addressing Biden in one sense at the debate, was not addressing Biden for the argument. The straw man, albeit in second-person grammatical form, was for an audience to see it in third-person form. His base, who likely would see the Senate's blocking Obama and Biden's appointments as a case of out-maneuvering the president and administration, beating them at a game they have mastered and should win, for the sake of righteousness. They would see any point scored on Biden as an important and telling one. Or, if the audience does not know anything about this background, they will see this as particularly telling evidence that Biden was not up to the task of being vice president—leaving that many judicial positions unfilled would be the height of administrative negligence.

The important takeaway is that there are many instances of addressing straw man arguments in the grammatically second person that are nevertheless rhetorically in the third. They are play-acting at addressing their dialectical opponents directly, this speaker to this interlocutor, since they are formulated with "you" as the grammatical form of address. But the argument, really, is not for that person, as its preferred audience is an onlooking third party for whom the picture painted will seem appropriate, that is, those operating under the hermeneutics of antipathy. The result is that there are, then, instances of straw man arguments *grammatically* in the second person that are *rhetorically* in the third person, as the speaker addresses the target for straw manning as "you," but only for an audience he intends to convince to say of the exchange that the speaker truly represented the target's vices and intellectual bankruptcy. It is, in the end, only dialectical theater.

5.2 The Effectiveness Puzzle, Strengthened

So far, our strategy for answering the effectiveness puzzle has been to say that straw manning works as a rhetorical device when it doesn't really address the target of the straw man but an onlooking audience either antecedently ill-disposed toward or not particularly knowledgeable about the target of the straw man. And we think this *third person* model of straw manning is a useful explanation for a wide variety of cases, even ones (as shown with "Judges") where it seems a speaker is directly addressing the straw man target. In short, straw manning in the third person isn't about convincing the target of anything but one's preferred audience. This strategy of answering the effectiveness puzzle is good to an extent, as it, we think, effectively handles what seems strange in straw manning in what seemed a dyadic dialogical sense—we've simply denied the dyad and argued that these dialogues are polyadic, with multiple audiences, and the argument was designed to be persuasive with them, not the target. And further, we think it is able to explain why instances of straw manning often seem so obviously shoddy from the perspective of their targets and those sympathetic with them—they are out to make a performance for an audience ready to use the hermeneutics of antipathy. But, we admit, this strategy has only *moved the problem*, as we have re-described the dialectical state of play from how we'd initially conceived the puzzle of effectiveness. The puzzle was how could straw manning be effective *for the target*, but our answer was that straw manning (in the third person) isn't straw manning the target *for the target* but *for an onlooking audience*. So, really, we changed the question. That's good in one way, since we clarify what the question was really about (perhaps a version of salutary straw manning, itself), but we nevertheless leave the initial puzzle unanswered. But, at least we now can pose it with more precision: *How can straw manning in the second person be effective?*

This strengthened version of the effectiveness puzzle will not admit of looking around for someone else to address the straw man to. Rather, it was what was behind our initial statement of the effectiveness puzzle in Chapter 1: "You won't think you've been refuted, because the reasons refuted weren't yours." So, now, to stipulate, the strengthened effectiveness puzzle is how straw manning is capable of dialectical success in argumentative contexts that are *dyadic* and the distorted, weaker, representation of the target of the straw man is presented to that same target as its audience. How can a straw man of you work on you?

Clearly, the strengthened effectiveness puzzle, to be answered, needs different conceptual tools. Strictly second-personal straw manning needs explanatory models that function either (a) internal to the dyadic dialogue or (b) internal to the target of the straw man. Our proposal, then, is that the functions of second-personal straw manning are directed at either (a) dialogical results (what we will call *silencing* output) or (b) at target-internal results of negative self-assessment (and so, are results of what is best known as *gaslighting*). In short, strictly second-personal straw manning is either out (a) to *silence* or (b) to *gaslight* the target with the negative misrepresentation. So, now, how does that work?

The silencing function of straw man arguments can be appreciated along the following epistemetric lines. Critical conversation is costly in time and energy, and if a particular conversation promises to be more costly than the results at stake are valuable, it is practically rational to not engage in the conversation and cede the results to the opponent. The picture of Pyrrhic victories (i.e., wherein the victor says that *if we win more battles like this, we'll lose the war*) is familiar enough (see Cohen 2005; Paglieri 2013). And, perhaps, we are all familiar enough with the colleague who gets his way consistently not because he is so skilled at argument and is reliably manifestly right about all things but rather because the transaction costs with him are so high, it's just better to just let him talk big and have his way. *Ad hominem* abusive works this way in the strictly second-personal instances, as the objective is to hurt the opponent's feelings in a way that makes continuing the conversation more emotionally taxing on them than it is worth. Calling people names doesn't change their minds; it just chases them out of the deliberative space. So the target of abuse retreats and the abuser holds the ground and wins the argument, at least in the sense that the abuser's view is the only one left represented in the dialogue.

Assuming that critical dialogue is a decision-making or inquiry-conducting game wherein the viewpoint to be endorsed at the close of the game is the one that survives all the rounds of the game's exchange, strategic maneuvering in the game is to make moves so that competing viewpoints do not survive and must be retracted. So some fallacious contributions to critical dialogues are those that impel a discussant to retract a view on the basis of issues irrelevant to the matter at hand. *Ad hominem*, again, works this way, as the discussant retracts from the discussion not because the view the discussant represents was criticized but because they do not want to continue to suffer more abuse. *Ad baculum*, too, is a way of forcing another's contributions out of the dialogue for reasons bearing directly on their personal well-being, not the epistemic standing of the view. They leave the scene, do not represent the view, and so, downstream, it has no defenses, development, or uptake (Casey 2020b). The same, we think, can go for the straw man in the second person. In this case, the speaker provides a negative misrepresentation of the target's view to the target, and the target then reasons that the work to correct the misrepresentation and get back to the issue at hand would take more time than they would prefer, and so they allow the representation to stand and either continue the argument saddled with the misrepresentation or bow out of the argument and cede it to the speaker. Consider the following.

5.2.1 Lottery Paradox

> Neela: The Lottery Paradox is serious business! Let's say you know that, with a fair lottery, one and only one ticket will win. But now, let's say you look at some random ticket—given the odds against it winning, you're justified in believing it won't win. But that reasoning is true for *every ticket*, so you're justified in believing that *no ticket will win*. Paradox!
>
> Aarnav: Ugh! That's not a paradox. It's confusion. Look—if you *know* that one ticket will win, you *can't be justified* in thinking none will. Seriously, what's wrong with analytic epistemology?

The problem here is that Aarnav misrepresents what the paradox is, since the paradox requires we think through *two* lines of thought yielding inconsistent outcomes, not one that yields the rejection of the other's conclusion. In essence, he states the paradox as a mere contradiction, and so a confusion. But that is to fail to understand the second arc of reasoning of the paradox, and so it is a misrepresentation of Neela's point—that, of course, it's a contradiction, but the reasoning behind yielding it is very good on both sides. That's how paradoxes work. So Aarnav straw mans Neela's view, and by extension, Neela and her philosophical training are also impugned. Neela, here, has two options: correct Aarnav about the structure of the paradox, that it has *two* lines of reasons internal to it, or just leave it and let Aarnav have this one. If the former, it requires that Neela will likely have to teach Aarnav about the concept of fallible justification, conjunction aggregation as a deductively valid inference, and the quantifier negation exchange rule to generalize from "all not" to "none." But, given Aarnav's expressed attitudes generally, this is going to be rough sledding—he's working under the hermeneutics of antipathy, and it's backed by some philosophical training that cements it. Alternately, Neela may just say that Aarnav can have this exchange, continue with his belief that a core issue of the analysis of justification is a simple confusion, and feel justified in his expressed antipathy for "analytic epistemology." It may just be too costly in terms of her time and patience to fix this, especially given that he's pretty clearly operating under conditions of the hermeneutics of antipathy that will make it so that whatever else she says will have similar philosophical yuckface. So she'll let him get back to whatever the heck he's interested in.

The point here is that Neela has not continued the dialogue, and so the straw man of her view *silenced* her. She does not correct Aarnav's depiction or defend the viewpoint. So, on the game model of the critical dialogue, Aarnav has *won*. His is the view left standing at the close of the exchange. But he has achieved this not by showing that there is something profoundly confused about Neela's view but that by showing his confusion about the view was so profound, it was too much trouble to fix it. This is a win for Aarnav in a purely dialectical sense, but it's a loss for all those involved. And this explains why second-personal straw manning can be effective but fallacious as a *silencing* strategy—it yields a dialectical result that is an illusion of the epistemic result of properly run exchange, and it misrepresents the resolution as one that both agree on the result and the issue is resolved. But it is inappropriately closed.

It also is worth noting that our explanatory model here is posited on the working assumption that in critical dialogue, silence upon being presented with a viewpoint, argument, or criticism is functionally equivalent with acceptance or assent. So, if an audience is presented with a claim amid an argument, if they do not register an objection or request backing, they have, for the purposes of the discussion, accepted the content, and it can be added to the sore of viewpoints from which to reason further (see Goldberg 2018 for a full defense of this view). However, the point of our line of argument here is to distinguish between silence as a *dialectical* contribution and it being a signal of other relevant considerations. Silence can sometimes be deafening in terms of rejection—jokes, for example, need something *other than silence* for them to be taken as successful. Silence can signal outright rejection, contempt, or it can be the result of nonargumentative strategies of ensuring it, such as understanding that objecting will yield untoward personal consequences (as noted by Dotson 2011; Tanesini 2018). But those features make sense only if we've seen the difference between what we've called being merely dialectically successful and what's a real argumentative success. So insofar as that default rule for interpreting dialogue results is right, we have a way of explaining *both* the dialectical success of the silencing versions of second-person straw man arguments *and* the fallaciousness of their usage—not only do they misrepresent the contributions of their targets but that misrepresentation and its consequent silence inappropriately close the critical exchange.

The defining feature of this silencing version of second-personal straw man is that we distinguish between the *dialectical* effectiveness of the strategy and its *epistemic-doxastic* effectiveness. That is, we can say that the target may *concede* the representation but not *believe it*, or we can say that it is nevertheless not recognized by the target as an *advancement* of our best reasons. So the straw man can still be a case of winning an argument (in a deficient sense, again), but not convincing the target or making a better case. It, in the end, turned argument into something less an exercise of intellectual virtue but more a battle of wills. But once we've stated things this way, a familiar problem returns. This, again, seems to be another version of our pattern of changing the effectiveness puzzle to make it easier to solve. First, we distinguished between the *onlooking audience* and the *target* for the straw man (the third- and second-personal distinction), and now with the second-personal version, we've distinguished between *winning* the dialectical exchange with the straw man target and *convincing* them of the move's effectiveness and the *advancement* of a better reason. Again, what we've done is precisify the puzzle of effectiveness by addressing the more tractable versions. But the newly precisified version of the puzzle of effectiveness is a particularly difficult one: *How can second-personal versions of the straw man not just win dialectically but yield a result of conviction from the target?* Is it really possible for someone to straw man *you*, to your face, and for you to believe it and for you to thereby change your mind? That's the strongest version of the effectiveness puzzle.

Our first thought is that, for the most part, the straw man in the second person is not effective in this strongest sense. Most folks smell the bullshit. It's because we generally know what we're up to when we argue—if we can identify the beliefs that divide us and then decide to argue, and if we've got the capacity to pick out reasons that support our views, we can distinguish between our reasons that are ours and ones that are not. In a way, though, this is just restating the effectiveness puzzle. It just doesn't seem possible, given what argument is. So, straw man, in the strict second-personal sense, is just an illusion? It doesn't really convince those whom you straw man, right? Well, things are a little more complicated, since there is a curious phenomenon to consider. Let's start with the fact that we are epistemically dependent beings. The straw man in the third person exploits that feature, again, because the straw manner constructs a representation of a target and their reasons for an audience. And it can be particularly effective if the audience is not antecedently familiar with the views of the target for straw manning. So, in hearing one side's representation of the other side of the debate, the audience members take themselves to know both sides, but it's an illusion. That's, again, the easiest version of the fallacy and the explanation of its effectiveness in the third person. In the second-person, that epistemic dependency is eliminated, but only in a restricted way. One reason it is not fully eliminated is that, though we know in argument what we said, our assessments of the quality of our arguments and views do indirectly depend on their uptake. That is, if you kept giving arguments that seemed to you to be perfectly reasonable but your fellows consistently reacted negatively to them, you'd start do doubt your judgment of their quality. The sensitivity of many of our normative concepts has this uptake issue. For example, ask anyone who values good jokes about problems with "funny" and audiences that resist laughing. Tough crowds get into comics' heads. Telling yourself that it was a good joke after stone-faced reception works only once or twice before you start asking yourself whether you know what's funny. Pandemic pedagogy is another case. Teaching to a Zoom screen of blank squares has unique challenges, since it's hard to tell if one's explanations are clear—nobody's nodding their head; there's not a moment when a student's eyes get big. There's just squares and names, and teachers start asking themselves if they know how to do it right. We need the proper feedback with these transactional exercises; else we start to doubt whether we know how to do it. The same, we think, can go for argument assessment, even of your own reasons. What's necessary here, then, is to introduce a new explanatory concept.

The phenomenon of *gaslighting* is the most theoretically robust way to capture what's happening in second-person straw man cases. Kate Abramson identifies gaslighting as a series of manipulations that induce in a target "the sense that her reactions, perceptions, memories, and/or beliefs are not just mistaken, but utterly without grounds" (2013: 2). The gaslighter aims at "getting another to not take herself seriously as an interlocutor." The term "gaslighting" is derived as a metaphor from the movie *Gaslight*. There is a couple, and the husband, Gregory, tries to make his wife, Paula, believe that she is losing her mind. He is doing so in order to take possession of her jewels. Gregory stages a discovery in front of Paula's friends that implicates she has been stealing from him, and he does so to make the case not only to others that she is unstable, but also to her. He searches for her jewels in the attic and dims the gas lights elsewhere in the house, and when she asks about the lighting changes, he tells her that these are hallucinations. So signs of her worsening mental illness. Hence, *gaslighting*. Again, the objective is to make a case to another that their judgment in some domain or in general is not reliable, and so it is deployed as a cumulative case, one that establishes a track record of contested results. And so, extending Abramson's account, we include *argument quality* as part of what is undermined.

The gaslighter generally takes advantage of either a power or perceived epistemic asymmetry between the two interlocutors, and it will also key on sexist, racist, ableist, or ageist stereotypes of the target for gaslighting.[7] So, in the movie case, Gregory's responses to Paula's judgments were to portray them both as symptoms of mental illness and as a particularly feminine weakness—in particular, that much of it is a result of her being "tired." However, there's nothing *essentially* sexist or racist about gaslighting. It's clear that gaslighting is easier in those instances, because the cumulative case can have more familiar uptake. But it need not depend on those. A case of simple epistemic asymmetry can show this. Consider, again, our example of "Hamlet's Ghost." In our example, Professor Elba and Lydia talk about whether the ghost in *Hamlet* should be on-stage or not, and Elba straw mans Lydia's proposal that it should by saying it would require someone in a bed sheet stomping around on the stage. We'd argued that the case can go many ways. If posed appropriately, with trust and expectation, Elba can give Lydia a chance to respond with a better staging idea or to note that *Hamlet* stagings can take many forms, and some risks can be worth taking. (We'll note that one of us witnessed a staging of *Hamlet* wherein Hamlet would intermittently wear Groucho-Marx-style fake glasses with plastic big nose and bushy mustache for delivering every humorous line, and *it worked!*) So Elba's representational second-person straw man can be an invitation, what we'd called a salutary or non-fallacious version of the straw man. It's a risky pedagogical move, for sure, but that's what a classroom of trust and positivity can allow—it's a gentle ribbing that can encourage heightened attention and innovation. Good teachers can do that, as can good colleagues. But without that background, or microculture, of trust and positivity, such a move can backfire significantly. Consider that Elba is a professor, older, more experienced, and established. And Lydia, by hypothesis, is a student, and so just getting used to thinking things through. She has no track record, and has taken the class so that she can come to understand these things. So Elba's representation of her view can give her the feedback that she's just not catching on, that she's not got what it takes. That her ideas are laughable. That kind of takeaway is all too easy for beginners to discussions, and it's why we noted that "Hamlet's Ghost" can be salutary, but it depends on the <ahem> stage-setting for it to go right. Without things arranged properly, the opposite result can ensue.

Abramson recounts Simone de Beauvoir's experience working out philosophically hard problems with J. P. Sartre. de Beauvoir argued for what she'd called a "pluralist morality," which Sartre criticized mercilessly and "ripped to shreds."

> I struggled with him for three hours. In the end I had to admit I was beaten; besides, I had realized, in the course of our discussion, that many of my opinions were based only on prejudice, bad faith or thoughtlessness, that my reasoning was shaking and my ideas confused.
>
> (quoted from Abramson 2013: 4)

We aren't given the details of Sartre's criticisms or their actual accuracy of representation, but the important thing to see is that a student's (or junior colleague's) self-assessment as competent at the task of arguing or reasoning can be significantly influenced by the assessments of teachers (or those senior). How we assess our reasons is influenced by how they are received by others, and a track record of rejection makes our own assessments less secure. This is the result of gaslighting, and being straw manned to your face, over time, especially if there's an asymmetry, can have significant effects on your own self-assessments.

The gaslighter is out to quash the source of objections and disagreement. Abramson calls it "radically undermining" the perspective of any deviation from the dominant perspective. So it attacks the *standing* of the target by turning their own assessments of their competence against themselves. The "moral horror" of it all is that, in the end, gaslighting "makes one complicit in one's one destruction" (2013: 17). And as Cynthia Stark observes, the *epistemic* force of gaslighting has the targets downgrade themselves as knowers on the basis of challenges to their credibility and reliability (2019: 231). Further, given that gaslighting will often be amplified by stereotypes of targeted groups, it functions not only as undercutting the individual target addressed, it communicates to all onlooking others of that group identity that the same awaits them if they venture into these critical territories. They are not welcome, and they are not able.[8] So there is not just a target in the *second-person singular* for the targeted speaker who is straw manned with these strategies but there is a target in the *second-person plural* for those of the same identity group as the speaker targeted.

To see this broadening function of second-person forms of the straw man, that of extending from *you* (singular) being negatively represented to *you* (plural), simply take this thought back to the educational context of "Hamlet's Ghost" and add to the example that Elba identifies Lydia's gender as a salient issue— perhaps he adds that "women can't help but be romantic about the supernatural." Lydia is being targeted with a straw man and with one that now highlights her gender, and this can communicate not just a negative assessment of Lydia but communicate a negative assessment of women's role in these discussions.

A further case of gaslighting wherein social identity is not relevant in the gaslighting message may be useful. One context for eliminating it in the communication may that it is under conditions of anonymous review for journal articles. So, consider the following.

5.2.2 Journal Review

It is a common enough occurrence that one's referees at the professional journals are not particularly fair. There are many contributing factors for this.

Journals ask for reviews to be turned quickly, they don't compensate reviewers for their time, and often reviewers agree to take on reviews without knowledge of how much work the review will take. (Aikin, for example, agreed to turn a quick review for an editor, only to find that the paper was fifty-two pages.) So reviewers are short on time and so can't put in the attention for the review that the authors have put in to producing the papers. But then add in the fact that there's often turf-guarding involved in the review process. A reviewer can have conflicted motives because they want *their* work on this issue to be coming out in this journal, and voting for acceptance may decrease the chance that their work will be accepted. Or the reviewers see their own work at the center of the field, and if an author's piece doesn't cite them (or cite them enough), that may be a reason to treat a paper, and by extension, an author, unkindly. And then there is the fact that most articles are reviewed anonymously, and the reviews are passed along anonymously, too. This gives the reviewers a bit of untoward freedom to be nasty to views they don't like without having to answer for it at the next academic conference. The objective of anonymous review is that the identity of the author (and reviewer) will allow an honest evaluation of the work in question, and though there are senses that this has trouble with conservatism with ideas, it does eliminate the worry that unacceptable prejudices in the first instance (of looking at a name, an affiliation, or title) would influence decisions. Ideally, reviewers read the papers without any knowledge of who wrote them, but it's not too difficult in small fields for a knowledgeable reader to figure out who's behind a paper. And if you have a distinctive writing style, anonymity is pretty hard to maintain. But for the most part, we can't make out who the authors are when we produce reviews, and that's the relevant consideration.

Now, let's imagine a graduate student, Ada. She writes a paper on a quickly developing controversial issue. The field is reasonably small, and since the issue is just about to crest, she's had the good fortune of timing to have been able to read all the relevant papers by all the players in the debate. And she's under the tutelage of one of the leading figures in the conversation, Professor Arglebargle. Ada, after some revisions on her own, with other grad students who know a bit about the area, and then with Professor Arglebargle, submits her paper to a fancy journal for consideration. Now, because Ada works with Arglebargle and they agree on many matters bearing on this issue, her paper has a number of references to Arglebargle's work. You know, Arglebargle (2016, 2018a, 2018b) and Arglebargle and Tweedcoat (2020) type references. And Ada takes an Arglebargle-style approach to the issue. It's not that Arglebargle has made an academic copy of herself, mind you. It's just that Ada and Professor Arglebargle think alike and Ada is keen to press this line of thought. Ada is, well, excited to wade in and mix it up, and Arglebargle is supportive. So off that paper goes! After being properly anonymized, of course.

A few months later, the referee reports are in. They are uniformly negative. Reviewer 1 (R1) points to some problems, issues that Ada knows are just standing objections in the area to Ada's view but ones she was pretty sure she'd answered. But R1 doesn't even talk about the answers—just repeats the objections, like nothing got said back. That's weird. And then says that the paper is "undertheorized and underdeveloped." But Ada thinks she can fix that, since it just means she needs to add another section to address this problem or that and announce it's a more modest project than Reviewer 1 thinks it is. But then there is Reviewer 2 (R2). This reviewer is clearly up on the literature and is a stakeholder in the debate. And R2 has some spicy things to say about Professor Arglebargle's views and arguments. And plenty more where that came from for folks sympathetic with that program. So the standard objections are trotted out as though they are utterly decisive and the paper's defenses are dismissed as rationalizations or *post-hoccery*. Reviewer 2 then turns to note how the author does not define the terms of the debate. Reviewer 2 notes that some assumption running the argument is controversial, so the argument can't go through at all. Reviewer 2 makes the case that the paper's examples are without context and too flimsy to build much theory around. They are "problematic," because they are really exceptions that prove the rule of the alternative view. Reviewer 2 asks why the author ignores the insights of the real stage-setters for the debate, Professors Y and Z (who are critical of Arglebargle and Ada's view, of course). Reviewer 2 says that the author doesn't even get Arglebargle's view right, because responsible scholars know that Arglebargle is motivated by other matters than what's at issue in this debate. And on it goes. Reviewer 2 has a very long and unrelentingly picky review. And then votes for rejection.

Now, at no point in the review is the author's social identity made salient in the review. For sure, it's salient in how it's received. If Professor Arglebargle got this review, she'd say, "Ha! Reviewer 2 strikes again! Well, I'll save this hot mess of intellectual hatchetry for a story over drinks later." And Arglebargle takes the paper elsewhere and gets on with her academic life. Haters gotta hate, you know, and R2 2 can't keep you from taking the paper to another journal, publishing it, and thanking them in a footnote later on. But it wasn't Professor Arglebargle who got the review. It was Ada. That's a very different story, and here, we can see how the straw man of Ada's view can be a blow to her own assessments of her abilities. A review like this will undercut her confidence in keeping track of the debates, it will make her unsure of how she uses the terms of the discussion, and it will undercut even her confidence that Professor Arglebargle sincerely approves of the work. Maybe Arglebargle is just being nice. A straw man like that for Ada is intellectually debilitating. And given the seeds of doubt planted by it, Professor Arglebargle's recommendations for how to deal with a negative review and later pep talks for more work and submitting the paper elsewhere ring hollow.

Gaslighting doesn't have to be intentional or make a social identity salient in the message. It is entirely a function of our epistemic dependence on others for our own self-assessments. We are fallible and social creatures, and that sociality can be a corrective to that fallibility. But that corrective function can overrun the self-confidence we need to have to reason and contribute to the social practice of exchanging reasons. Reviewer 2 wrote their review out of a maximally adversarial comportment, interpreted the paper thought a lens of the hermeneutics of antipathy, and produced a straw man of the author's view. Reviewer 2 had no intention to gaslight anyone but just wanted to give a smackdown to a view they revile. And even though Ada previously had a grasp of her argument, how the field is laid out, and how it all goes, this straw man of her case given back to her undercut it all. And so, that's not just how straw man arguments can be effective in the second-personal form but that's also how a gaslighting message needn't invoke the target's social identity to erode their self-appraisal as epistemically reliable.

The strengthened version of the puzzle of effectiveness for straw man arguments is that it seems difficult to see how a negative misrepresentation of a speaker's view to the speaker could be argumentatively effective, because the speaker will be able to keep track of their views and detect the qualitative difference between the given view or argument and the representation. Most other cases of straw manning, in the third person or strictly as dialogically cynical strategies, are posited on the attitudes of the target of straw manning being beside the point—they are not out to change the target's mind but to sideline them in the conversation. But in the case of some instances of second-personal straw manning, it does seem possible to change the target's view. It is by making a case to tell the difference between a better or worse version of their view, by undercutting their own self-assessments of competence of argument. So our answer to the strengthened version of the effectiveness puzzle is to say that these cases of straw manning, if effective, are instances of epistemic gaslighting. And we will hasten to add that though this form of argument can be purely dyadic (with just a speaker and target), the polyadic versions (with speaker, target, and onlooking audiences) can create broader cultural antipathies and erode intellectual trust. Insofar as that is the case, straw men breed more straw men.

5.3 Taking Stock

The puzzle of effectiveness, as it turns out, is a pretty hard problem. How can the critique of a view you don't hold be effective in changing your mind on an issue? The answer required is that we precisify the problem, and as we become clearer about the problem, we see that there are three general classes of answers. The first is that straw manning in the *third* person, addressing the misrepresentation and critique to an onlooking audience either ignorant of the target or with shared antipathy for the view, explains how the argument can be effective, but not on the target for the straw man. In cases where it is important to a speaker to convince a crowd (often of their supporters) that another discussant is wrong, getting that discussant's view just right isn't quite as important. And it certainly works better if the target for misrepresentation isn't present to set things right. In cases of straw man in the *second* person, when the misrepresentation is presented to the target, the strategy can be effective in either *silencing* or *gaslighting* the target. Second-person silencing works in the fashion that many other fallacies work in closing off cooperative paths for productive dialogue. One overtly signals that one is just hard to deal with. With *ad hominem* abusive, one calls one's interlocutors names to chase them off, and so the straw man works in similar fashion—one interprets one's interlocutors in terrible fashion to make it prohibitively costly to continue the exchange. And so these interlocutors are *silenced* by this strategy. With the case of gaslighting, one takes advantage of the symmetric features of reason assessment with argument. Reason quality should have a public face, that of reason appreciation. So if one paints the reasons of another as not worthy of acceptance, there are instances wherein those receiving those messages about themselves and their reasons see this feedback as evidence that they themselves are not reliable judges in these matters. Argument requires that reasons be manifestly good, and so denying the manifestness is an undercutting case against that reason's quality. And so, one has *gaslighted* an opponent with a straw man of their view, and so yielded a change in their assessment of the quality of their case. These three explanatory stories, together, we think, answer the effectiveness puzzle.

6

The Puzzles of Dialecticality and Meta-argumentation

At the opening of this book, we posed three puzzles for a theory of the straw man fallacy. The puzzles of effectiveness, dialecticality, and meta-argumentation. The puzzle of effectiveness was: how can straw manning be effective given its core of misrepresentation of what is criticized? Our answer in the previous chapter was that the effectiveness of straw manning depends on the personal form of address for the arguments. If in the third person, the misrepresentation is either not detected by the audience or is, from their perspective, beside the point. If in the second person, the misrepresentation is either a strategy of turning the dialogue into a battle of attrition with the target and thereby silencing them or a form of gaslighting the target so that they do not trust their own judgment of reason quality. The two remaining puzzles come as a piece, since the straw man's core is the negative misrepresentation of another's argument and critique of that represented argument to close the critical dialogue. The dialectical puzzle is articulating just what is so wrong with criticizing a bad argument, and the meta-argumentative puzzle is that it is surprising that reasoning about reasoning yields a distinct set of fallacies, with straw man being the most prominent. The key to both, as we see it, is that these two puzzles about straw manning are good ways to capture an important point about argumentation generally—when we argue, we are out to make progress on something controversial, answer a question, address an objection. Arguments are public acts and products of our reasoning, and they not only are supposed to be rooted in our thinking together about the things we think about (the weather, dinner plans, economics, and the Peloponnesian War) but also to make it so that we endorse how we've thought about those things.[1] That is, arguments not only direct our attention to the things about which we reason but direct our attention to where a worry might be about the conclusion, and if rightly addressed, call attention not only to the quality of the conclusion but how we got there. The words *so, therefore*, and *hence* highlight that we've made an inference and implicate that we've satisfied the scrutiny that the question prompting the reasoning impels and that this should be enough to satisfy us. So arguments generally are dialectical and meta-argumentative—you pick up reasons that are appropriate for the controversy, accessible to those with whom you argue, and of quality for the purposes at hand. And when you finish, you announce your assessment of the job done well. *Therefore* is not just an endorsement of the conclusion but it is also an endorsement of the reasoning that yielded it and an assessment of it living up to the standards behind the need to argue in the first place. The straw man is where the dialectical and meta-argumentative features of argument are clearest with its fallaciousness, since it is the nature of this fallacy that it is clearly a dialogue error of arguing incorrectly about an argument.

6.1 Argument and Contrastive Reasons

It is worth pausing to note these features of dialecticality and meta-argumentation are in high relief when we see the contrastive role of reasons in argument. Contrastivism about reasons is the view that all reasons do the work they do indexed to a *contrast class*, so reasons function not just as reasons *for something* but rather as reasons *for something as opposed to something else*. A few examples may help make this point clear.

> **Practical Reasons:** Should we go to the Burger Hut for dinner? If I say, "Yes, because it's close," that may be a good reason if the alternative is the faraway Taco Palace, but it does not work if the alternative is the equally close Curry Cabin.
>
> **Moral Reasons:** Was Adrianna running into the burning building and saving Tiny Tim praiseworthy? If we say, "Yes, because she saved a life at risk of harm to herself," that reason holds so long as it was as opposed to her staying out of the fire altogether. But if Adrianna could have *also* saved Tiny Tom in the process without any more trouble, then it seems the praise is mitigated (from Snedegar 2015).
>
> **Explanatory Reasons:** Why is the sky blue? If we say, "Because the sun is up," that explains why the sky is blue instead of dark (as in, at night). If we say, "Because blue light's shorter wavelength makes it more diffused in air than yellow, red, or green," then we explain why it's that color as opposed to those (from Sinnott-Armstrong 2008).
>
> **Epistemic Reasons:** How do I know you are having iced tea with lunch? If my reason is that what I see in your glass is a light-brown and non-carbonated

beverage with lots of ice, then that's a justifying reason, assuming the other choices to drink with lunch are water, sodas, coffee, and so on. But not quite if we're in a bar and Long Island Iced Tea is on the menu.

The big idea here is that a central feature to evaluating the quality of a reason is in evaluating what it's a reason for from a set of alternatives. If we change the alternatives, the reason's quality also changes. What reasons do, then, is sort the best of the bunch out, and that (given the options) is what we should accept. Reasons evaluation, then, is *triadic*, between (i) the reason, (ii) what the reason favors, and (iii) the alternatives, or what's commonly called the *contrast class*.

There are a number of benefits from the contrastivist take on reasons. The first (as noted by Schaffer 2004) is that we have the tools with this program to explain what seemed mysterious about *contexts*—why does reason quality change when we change interests or with whom we are talking? Contrastivism explains that change in quality with the change in contrast classes—in some discussions the contrast class is different from others because some alternatives are salient from the perspective of one group but not with others, so the reasons themselves, to do the work of reasons, must be different, too.

A further benefit of contrastivism is that, in epistemology, it allows us to explain the appeal of various skeptical scenarios but also retain the idea that even against skeptical challenges, we can identify better and worse cognitive performances. For example, a person's kinesthetic and visual experiences of their hands may give them good reasons to believe that *they have hands rather than wings or flippers*, but it does not give them reason to believe that *they have hands rather than being deceived by an evil demon* (as noted by Baumann 2015; Dretske 2013; Sinnott-Armstrong 2004).

An important lesson of contrastivism, consequently, is that *issues* for critical evaluation are clarified by not only what is in question, but what the range of alternatives is. And so, if the question is whether John should, say, *ride his bicycle to campus today,* what reasons we can bring to bear on the issue depend on what the alternatives are. So if the issue is between bike or bus, that *John needs exercise* is a reason that performs contrasting work. But if the choice is between bike and walk, then that reason doesn't (from Snedegar 2013). The key is that before we start deliberating, it's best to identify our options so we can then aggregate and evaluate our reasons appropriately. And it's a familiar enough experience for many that, as deliberations proceed, new options may arise, and so returning to previously settled matters is necessary. Contrastivism explains why that's good policy—reasons that once did contrastive work now may not, given these new alternatives.

As we noted in Chapter 1, argument has an adversarial core, and contrastivism explains why that must be the case—reasons are reasons because they play a contrastive role of eliminating a relevant alternative. And this is why dialectical features of argument are so important, because we need to be clear on what the alternative considerations are for us to be able to judge reason quality. That is, we can evaluate reasons only in light of what the alternatives are that those reasons need to manage and eliminate, and in order to do that evaluation of those reasons, we must know the contrasts. So, in order to reason out an assessment of the quality of an argument, we must access the dialectical context providing the contrasts. We need a representation of what others are saying about the controversy to judge whether our reasoning is sufficient to the task of settling the issue. Without that representation, we have no way to judge whether the reason does appropriately contrastive work. An example may be useful to show this, so imagine that you are deliberating with colleagues about your department's class offerings for the next semester. You are deciding between adding an extra Composition course and another Critical Thinking course. Both are service classes for the university, and you have reason to believe that whichever you offer, it will fill. Given that background, if one of your colleagues insists that we need to offer the Composition class because it is a service course and will fill, we can agree that these are relevant considerations in a sense, but they are not cutting in favor of the writing class any more than they would for the logic class. Even though relevant in the sense that it does support accepting the class, it does so equally with its contrary, given the current options on the table. So, it does not do the appropriate *contrastive* work of a reason here. And since the reason supports *both*, and we can choose only one, it in the end supports *neither*. Note, importantly, that were we deliberating between the writing course and a special seminar on the metaphysics of mystics of later German scholasticism, what your colleague gave would be a good argument, because that metaphysics class is neither a service course nor would we expect it to fill. But because we are deciding between courses that are both service courses and expected to fill, it's not a good argument. Notice that to make the reasoning explicit required to note the need not only for a reason of relevance but a reason that does the salient contrastive work given what the alternatives are, we must go meta- to make explicit the standards for appropriate reasons for the context. And in doing so, we survey the contrast class the reasons must bear on. If the reasons bear equally on the proposition supported and the contrasts, it's not a particularly good reason. But if the contrast class is changed, that changes the quality of the reason. And to say *all that*, we need meta-arguments, and these meta-arguments require that we look at the contrast classes for evaluating these reasons.

The most robust and clear way to fill those contrast classes is by way of the disparate opinions of those with whom we disagree and deliberate. And so, not only is there a pragmatic reason for supporting the norms of dialecticality in argument, there is a reason here because of the contrastive nature of reasons as reasons. The upshot, then, is that dialecticality and meta-argumentation are internally connected, because argumentation is intrinsically contrastive, and this contrastivity is intrinsic to arguments as publicly endorsed lines of shared reasoning. That's a big thought, but it was worth nailing down before we address the two puzzles.

6.2 The Dialecticality Puzzle

You can't commit the straw man alone. It is a *takes-two-to-tango* fallacy. Yes, it may be possible to straw man yourself to yourself, but it would require that you do so to a past or future part of yourself (say, representing reasons you used to endorse as worse than they really are), but the heart of that is yet in the third person dialectically (as in, that the *old me* isn't really *me*, but someone with my name and a lot of my earlier memories). The straw man in dyadic form requires a *speaker* and *target* of representation, but in triadic or polyadic form, it takes a *speaker*, *target*, and *audience(s)*. That makes the straw man different from many other fallacies, since one most certainly can commit many of the rest of them on one's own. It does not take an active dialogue partner to make a hasty generalization or commit the fallacy of asserting the consequent. Those arguments, internal to their form, don't require that we overtly represent the thoughts and reasonings of another for evaluation. But the straw man does. It is, for better or worse, a social fallacy, one borne of our shared critical dialogue.

One feature of the sociality of the straw man that is the puzzling part is that it is *overt criticism* of what's been identified by another's bad reasoning. In the third-person form, the speaker attributes some flawed argument to the target, and the speaker and the speaker's audience reject that bad reasoning. In the second-person form, we assume that in both the dialogical silencing and gaslighting forms, the straw man also requires that the target (as audience, too) rejects the attributed reasoning as flawed. And so, importantly, the argument attributed is unproblematically bad and the speaker and audience agree on that evaluation, and they all reject it. So, we begin with the question: what's fallacious about rejecting bad reasoning or manifestly objectionable viewpoints? And it's here that anyone who says it's *irrelevant* misses the point. One reason why is that the question is *what exactly relevance is* under these conditions—surely, it is relevant in a semantic sense to show that a view or line of reasoning bearing on the question is false or unacceptable. That is, assuming that the negative representation presented is problematic as a view or argument, then rejecting it *is* relevant to the conclusion, as it, in fact, eliminates bad reasoning or false views from consideration. And, as we'd noted in Chapter 4, there are salutary versions of straw man in all its forms that are relevant to the dialectical context, even if they misrepresent the overall dialectical state of play. It's because they eliminate problematic views and lines of thought that may be tempting to those in the discussion. And so, if the answer to the problem of dialecticality is *relevance*, we have a significant problem.

Now, an appealing second answer is that the issue with straw manning is not with the rejection of the flawed argument, but with the false attribution of the argument. It's an inaccurate representation of the target's views. So it is a dialectical misfire. That is a very appealing answer, and many take it that this is the end of the story. But, as we've argued, misrepresenting the other side's views isn't always fallacious, either. Given that there are argumentatively salutary straw man arguments, it can't be the *misrepresentation* that explains the fallaciousness of the straw man. And so, now that we've looked at ways the dialecticality puzzle hasn't been properly addressed, we can see this as the *hard version* of the dialecticality puzzle—one does not get to invoke relevance or misrepresentation to explain the fallaciousness of straw men. Now, that's a hard problem. An example may sharpen this issue.

6.2.1 INFOWars Comedy Hour

Nutpicking, or weak manning one's opponent, is a form of the straw man fallacy wherein one finds the worst or weakest version of your opponent's views or the least sophisticated defenders of an opposed view and then subject that view to high scrutiny. So one goes after the bad versions of one's opposition, instead of the good ones. The strategy can occur in lots of ways. One can wait for an offhand and awkward comment to encapsulate the view, or one can track down the least informed representative of the opposition. Or one can listen in on the other side's loose talk. This last one is a new way to weak man, especially given that loose talk that would otherwise be private is now broadcast as entertainment for the loose talk in-group—so, one listens in on a venue of entertainment by and for one's opponents and wait for them to say something that sounds all-too-revealing about their objectionable views and their otherwise hidden vices. Reporters at INFOWARS did just that. They listened in on comedian Michelle Wolf's political comedy Netflix show. In the months before President Trump's attempts to work out a nuclear disarmament deal with North Korea in 2018, Wolf devoted a full show to commenting on what she saw as the absurd situation that the fate of the world hung in the balance depending on whether Donald Trump and Kim Jong Un could get along. Already comedy gold, if you appreciate the dark stuff. And in a setup bit, Wolf asks her reliably left-leaning and Trump-despising audience the following question:

> Are you sort of hoping we don't get peace with North Korea so you won't have to give Trump credit?

A funny question. Of course it's a joke, but one that is at the expense of admitting the deep resentments at the heart of American politics. The joke gets funnier, since the audience polled answered Yes 71 percent to No 21 percent. They played along with the bit! What a lucky comedian! Well, it's all pretty funny, and surely everyone who responded had a little chuckle. As did all the liberal-leaning audience members at home. It's loose talk, and they were having a laugh. Oh, but the INFOWars folks were listening, too. They don't like humor, unless it's them making a joke about how sensitive liberals are. Anyway, Paul Joseph Watson, the INFOWars author, didn't get the joke, and then reported:

> In other words, a significant majority of leftists would happily risk nuclear war, so long as it meant Trump would look bad.
>
> Let that sink in.
>
> When conservatives talk about how many on the left "hate America," it's seen by most as a tired cliché, but when you see clips like this it really makes you wonder....
>
> Indeed, it seems that the left is so beset by Trump Derangement Syndrome that they're quite happy to see the pilot crash the plane even though they're on it.[2]

So here is what the INFOWars article comes to: a reporter watches a comedy show and reports that a gag that the audience was supposed to play along with bespeaks a traitorous vendetta among liberals. The comedy bit, apparently, was taken as reliable polling data, no less. So much of the straw man fallacy generally is about interpreting your opponent in a way that exercises minimal charity, if only for the sake of the quality of the exchange that these defaults encourage. And in this case, it both depends on and is evidence for the hermeneutics of antipathy. But, to be clear, if your defaults are set on interpreting a comedy sketch like this as little more than a suicidal desire for Trump to fail, then it's hard to see how there's much of any opportunity for critique either way. The comedy sketch, admittedly untoward, because it is a *comedy* sketch about people talking about their political preferences rooted in resentment, is supposed to be an instance of many laughing at themselves for what they themselves see as bad motives—they see themselves beset by Trump Derangement Syndrome, too! What makes it funny is that they admit it, but *laugh at the motives to nevertheless reject them*— it, of course, is absurd to hope for a nuclear war just so a person you despise cannot claim credit for preventing it. The laughter is not an endorsement of the attitude but an honest admission of its temptation and repudiation of it. Humor, just like human psychology, is complicated, but the laughter from those liberals was at themselves. And it's that part that the report about the comedy bit left out.

So the takeaway is that in "INFOWars Comedy," *everybody* rejects the view attributed, and the view attributed is manifestly bad. That's a kind of progress toward and maintenance of sanity. The problem, however, is that the view and attitude are falsely attributed to the liberals, and on the basis of this false attribution, further inferences are drawn about their moral and intellectual character. This is why the INFOWars reporter tells us to let this information "sink in" and turns to broader points about the left hating America and preferring to crash the plane that they are on—that they are beyond being reached by civically minded reasons or even appeals to their own self-interest. Notice that it's *here* that the problem of the straw man resides—it's not that simply a bad view was falsely attributed but that *on the basis of that bad attributed view, the audience is to infer that the target is not worth arguing with further*. The invitation to "let that sink in" is an invitation to this kind of meta-argument.

We've identified the core problem of straw manning (and iron manning) as its *closing function* on the arguments. That is, what distinguishes the salutary from the fallacious instances of the straw man is their *custodial* role on the ongoing or ending of the discussion. In the case of straw man arguments, bad views and reasoning get attributed to the target. In salutary cases, the garbage is thrown out by both parties. The speaker rejects the view attributed, and the target, after rejecting the view too, is allowed a clarified version or another shot at arguing. The critical conversation is improved by identifying the trash and cooperatively taking it out. And so, with "Moral Atheism" from Chapter 3, for example; the weaker arguments are selected as weak men for criticism not to portray the other side broadly and then to end the argument but for the sake of clearing out a regular confusion with the issue so that more focused discussion can be pursued later. The argument is not closed with the misrepresentation and its criticism, but it is opened for better arguments and clearer considerations. And iron manning clearly has such a salutary function, as we see with Professor Zoccolo's improvement of Alfredo's critique of Rawls and the Original Position in "Philosophy Student I." The iron man there opens the conversation to an on-ramp with broader communitarian alternatives to liberalism. However, when the misrepresentation *closes* the argument to further development with the target, we see the fallacy. So we see no hope presented for improved reasoning in "INFOWars Comedy," precisely because not only do we take away the attributed thought that the view deserves no more critical attention but the targets for representation do not deserve much, either.

The attitudes behind this interpretive result are those of what we've called the *hermeneutics of antipathy*, a set of ways of interpreting one's interlocutors in the least charitable lights, expecting intellectual and moral vice behind their statements, and being primed to respond with resolute refusal to listen any more to them. And, as we'd noted, this stance has a hermeneutic circle of support when in place—one interprets these targets with suspicion and antipathy, and one invariably finds what one was looking for, and so further adds evidence for the wisdom of maintaining this comportment of suspicion and contempt. And once one starts seeing straw men of one's opponents, they come in flocks. In such a case, one constructs an army of straw men to battle, often all at once, as a Leviathan of straw, or what we call the *Burning Man*. It is a unique kind of straw man perpetrated when one creates a pastiche of distortions of one's dialectical opponent—it is not composed simply of a single distortion but rather a slew of mischaracterizations bent on representing one's opponents in the worst light. In deploying the burning man fallacy, one not only stuffs an opposing figure with straw but then proceeds to surround it with more tinder and additional flammable material, with the intention of committing the view at issue to the flames, along with whole traditions, movements, and ways of thinking.

6.2.2 Meletus and Socrates

One famous example of the burning man is found in Meletus's exchange with Socrates in Plato's *Apology*. Meletus's charge is that Socrates is a representative of everything wrong with intellectual life in Athens, and so he is accused and stands trial for impiety and corrupting the young. According to Meletus and the older accusers, Socrates is an impious physicist who holds that the sun and moon are mere stones, that he makes the weaker argument the stronger, that he is not capable of raising children properly, that he intentionally corrupts those around him, and that he is an atheist. The problem, of course, is that Meletus has depicted Socrates as if he were Anaxagoras the physicist and that he is one of the immoralists and libertines of the sophistic tradition. Meletus has taken a collection of bad representatives of the intellectual climate in Athens and just thrown them all at Socrates. These are all bad things (by hypothesis), but none are accurate with respect to their target, and they are deployed not just for the sake of a criticism of one of Socrates' views, but for the sake of a full rejection of everything about him. It is less an attempt at conviction along evidential lines, but more through a repudiation of the *kind of person* that Meletus claims Socrates is. And it's worth noting that Meletus's case is in the (at least rhetorical) *third person*—it is about Socrates, to the jury. And it not only is a repudiation of Socrates's character and representation of him as an intellectually dishonest person but it concludes that the jury should beware of Socrates when he speaks. Meletus, assuming Socrates's reply reflects what came before, closes his initial speech with a meta-argument that looks not just at what Socrates *has done* but prepares his audience for what Socrates *will do*. This is why Socrates must open the *Apology* with his attempt to undo Meletus's warning that he is a *clever speaker*. It is not just about the fact that Socrates has argued so cleverly in the past to corrupt the young but that his coming arguments are going to be sources of corruption. The point overall is that Meletus collected a wide and potentially inconsistent set of representations of Socrates and threw them all out at him. Such a collection of straw men, again, is what we call a *burning man.*

Let's turn to a more current example. This tactic of burning the man is common when specific disagreements are tips of icebergs reflecting larger cultural divides. And once a disagreement at hand is seen as a manifestation of the cultural divide, all the bad images of the culture that one abhors now accumulate around one's dialectical opponent. It happens quickly, and it yields little more than expressions of revulsion with those on the other side. The hermeneutics of antipathy bleeds over into everything one sees the other side does.

6.2.3 Hannity's Democrats

Sean Hannity of Fox News is a master of the burning man. In the fall of 2019, in the run-up to the Democratic Presidential Primaries, he presented in one short ten-minute piece a characterization of the "crop of 2020 Democratic hopefuls" as collectively:

pushing to give 16-year-olds the right to vote, even though you can't drink until you're 21. They're promising to stack the Supreme Court so they get enough justices that think their way and will legislate from the bench. And they're proposing an end to the Electoral College.... Government-run health care and complete takeover of the health care industry—yes, run by the state.... And government-run education - now they're going to add pre-K and college education. They haven't done a better enough job with kindergarten through 12th grade? And they will have government consolidation of all guns and gun laws and government-run clean energy through the government takeover of the energy sector.... Government-run universal income, government-paid-for vacation, government-sponsored healthy food

... [T]hey will tell you how to run your business. There will be government promises for everything and for everybody—you'll never have a worry in the world. But you also will give up all of your freedom.

The strategy here is to collect a set of views—specifically those objectionable to his preferred audience of Fox News watchers—and pile them together into a heap of disagreeable ideas. His strategy is to show not just that any one of the commitments is wrong but that they come as a wholesale package of something not to be argued against, but simply reviled. He concludes that:

Democrats don't share the values of our framers. They want power for themselves, and they want it at all cost.... It is a blatant, dramatic, frightening attempt to alter America in ways that will make it unrecognizable and forever destroy the greatest economic wealth creation system in the history of the world.

It's pretty dramatic stuff, and the rhetorical strategy is to enable the aggregation of outrageous commitments to overwhelm. Elsewhere, Aikin and Talisse (2018) have termed the strategy of overwhelming an opponent or audience with ideas, arguments, objections, and proposals as *swamping*, and the burning man uses much of the same strategy. Instead of one particular line of higher-quality reasoning, the sheer quantity of claims, combined with the intensity of outrage that they provoke, takes on a quality of its own.

The burning man, then, not only distracts from the actual claims and arguments given by particular interlocutors, it's also a way of preventing oneself (and one's audience) from even hearing those claims and arguments in their nondistorted forms. In repudiating those people as if they were simply a collection of objectionable commitments, one forecloses further exchange with them.

With the burning man, the two general problems with straw man fallacies come into especially tight focus. Not only does one distort the picture of one's dialectical opponents, but one does so in a way that disables argument with them.

In representing them as a contrast class that requires only very simple and easy reasons to dismiss, one provides reason to no longer take them seriously. If we value a culture of critical exchange, both of these results must be resisted. But the burning man can be used as a salutary contribution to critical discussion, too, as it can function as an aggregator of the general problem with a proposal. And, again, this is the reason why the simple misrepresentation answer is insufficient to explain why straw man is fallacious.

6.2.4 Chinese Philosophy

Bryan Van Norden, James Monroe Taylor Chair in Philosophy at Vassar College and Chair Professor in Philosophy in the School of Philosophy at Wuhan University, was interviewed at *What Is It Like to Be a Philosopher?*[3] Professor Van Norden specializes in East Asian philosophy, particularly Confucianism and Buddhist metaphysics. He was asked about how hard it was to find a job teaching Chinese philosophy in Philosophy departments in the United States. He replied that though most philosophers recognize the depth of insight in Asian traditions, they resist calling it "philosophy," preferring to hold that there is no properly termed philosophy outside of the Greco-Roman tradition. Van Norden would find himself in arguments with his colleagues and other philosophers about this assumption and the question of expanding the canon. He found himself on the receiving end of a number of bad arguments, and he collected them. He recounts them in his interview as one (fictitious) conversation, of which are some selections:

> I have had versions of the following conversation more times than I care to remember:
> Me: Have you considered teaching Chinese philosophy in your department?
> Colleague: Philosophy is by definition the tradition that goes back to Greece.
> Me: That is not even a good *prima facie* argument. What makes something philosophy is its topics and methodology, not an accident of historical association. For example, mathematics exists independently of the Anglo-European tradition, so why shouldn't philosophy?
> Colleague: We don't teach religious studies or the history of ideas, only genuine philosophy.
> Me: What Chinese thinkers have you read that you believe are not really philosophers? Mozi? Zhuangzi? Mengzi? Xunzi? Han Feizi?
> Colleague: I haven't read any of them.
> Me: If you haven't read any of them, how do you know—
> Colleague: but they're all just aphorists.

Me: Heraclitus, Pascal, Nietzsche, and Wittgenstein are aphorists, and they are philosophers. Besides, most Chinese thinkers do not write in aphorisms. That is a stereotype.

Colleague: But they don't discuss the same philosophical topics in China that we do in the West.

Me: Yes, they do discuss many of the same issues, including topics in normative ethics, meta-ethics, epistemology, and metaphysics.

Colleague: Maybe they discuss the same topics, but they don't use a philosophical methodology. They don't provide arguments.

Me: Yes, they do. I'd be happy to give you a dozen examples off the top of my head.

Colleague: Why can't you just teach Chinese philosophy in areas studies or ethnic studies or something?

Me: Why can't you teach Kant in the German Department or Rawls in American Studies? Why do we even need a philosophy department instead of different area studies? The answer is that Chinese philosophers should be taught in philosophy departments because they are philosophers, and philosophers use distinctive approaches to teach texts that people in language and literature or area studies departments typically do not.

Colleague: We can barely cover all the figures and texts in Anglo-European philosophy now. What would you have us leave out?

Me: You are nowhere near close to covering all of Anglo-European philosophy now, you never were, and you never will be. It's always a matter of deciding priorities, and I have seen many departments will multiple specialists in the same field in Western philosophy but no one in any branch of non-Western philosophy.

Colleague: So you think everything in the West is bad?

Me: I never said any such thing. In fact, when it comes to epistemology I am a Neo-Kantian.

Colleague: But you think Chinese philosophy is better than Western philosophy?

Me: I didn't say that either. I value them both.

Colleague: Look this is the tradition we work in. Take it or leave it. [Note: Yes, someone I know was told this in response to the suggestion that they add non-Western philosophy to the curriculum.]

Me: Have you ever heard the expression "the unexamined life is not worth living for a human"?

What we think is important to note about "Chinese Philosophy" is that it is an instance of *burning man*—Van Norden has collected a group of weak men for a full argumentative drubbing. But it does not have the closing function we see from that with "Hannity's Democrats." The difference is that Van Norden's strategy is one that does not close off improved conversation on the issue. It, like "Religion and Morality" in Chapter 3, functions as a clearing of the decks for better conversation—once we just get past the common errors in the domain of discussion, we can have a fulsome exchange. And it may be that, once we see the common errors and their extent, there may be no other resistance. But regardless of that result occurring or not, the cattle call of bad arguments functions as a list of frequently committed errors to avoid than evidence that the target for representation is beyond argumentative help. (Again, it is a form of weak men that *actually have been given*.) In contrast, "Hannity's Democrats" is posited on the thought that the aggregation of depictions is yet more evidence for why the Republican Party must mobilize for a coming battle with vicious and intellectually dishonest opponents.

The dialecticality puzzle for a theory of straw man arguments was that criticizing bad arguments is a good thing, and misrepresentation of an opponent's view is itself not necessarily vicious (since there can be salutary straw man arguments). So wherein does the vice of fallacious straw manning reside? Our answer is that it is in the premature closure of the argument. The negative misrepresentations, when produced from and contributing to the hermeneutics of antipathy, not only close critical discussion of the topic at issue with those interlocutors, it casts them as those with whom one cannot have discussions of quality. It is, as we argued in Chapter 3, *the closing function* of these arguments that distinguish fallacious from salutary instances of straw manning.

6.3 The Meta-argumentative Puzzle

Every year, both of us have conversations with university administrators, colleagues in our department, chairs of other departments, and the occasional university donor about the logic and critical thinking classes we teach. We generally have enthusiastic audiences for our cases that the university should support teaching these classes, and perhaps that we should offer more. Compared to making the case for so many other classes taught in our philosophy departments (say, Ancient Greek Philosophy, German Idealism, or Aesthetics), this is a very easy sell. We aren't out to down on those other areas. In fact, some of our best friends are Hegelians. Rather, our point is about comparative uptake for the logic classes. The reason why is a reasonable default with many, and it's one we key on when selling the class. Learning how to reason about reasoning, developing the skills of evaluating arguments, will teach us to reason better. That's the assumption behind it all. And it, again, isn't unreasonable. The language of logic and argumentation is a window through which we may take in our reasoning, so in learning the language of logic we may self-consciously reason. And when we say we are able to reason reflectively, the promise of improved reasoning seems manifest. That's the promise, and that's what we tell the deans. And they believe it, no problem. (But there's a secret that we'll tell you shortly that you can't tell the deans: it's about how learning all this stuff makes for *whole new fallacies*.)

Meta-argumentation is argument about argument. Most of our arguments are about the things that unreflectively occupy us, so we reason in the first instance about our cats, our bank balances, whether it's Finland or Norway that has the famous fjords, or whether driving over the speed limit at this time of day will risk a ticket. Our first-order reasoning is regularly in an object-language, with terms that point beyond our reasoning and our language. But then we say words like *therefore* and *so*. Those words don't point to the things we reason about, *but to the reasoning, as it is coming to be completed.* And so, we, even as we reason about the things we reason about, we also come to think about our reasoning *as reasoning.* We endorse what we did not because we just think we got it substantively right about the cats, the bank statement, or that it's definitely Norway with the fjords but that the procedure we used to get there was right, regardless of what we were reasoning about. And so, we are drawn to reason about that reasoning, seeing why it was well-done. An inference along the following lines may occasion little explicit argument about it as reasoning, but we are able to bring it out with our meta-language of logic:

> The cat is howling by her food bowl; and when she does that, she's hungry. So she must be hungry.

That's all reasonable enough, but we can see a pattern that has nothing to do with cats or their food, but about the form that the reasoning takes. We identify *premises*, *conclusions*, and even the *inference forms*, and we evaluate the complexes as *valid* or not. That's taking the meta-level view, and when we think our way through this publicly, we give *meta-arguments*. We argue about the quality of our arguments. And, again, selling the skills of all this is very easy to the deans and chairs of academic departments in the humanities, since it's appealing on its face—that developing the skills and attitudes of reasoning about reasoning should show us better results. That's *Geisteswissenschaften* if there ever was such a thing, folks, and it is why we have enthusiastic deans in our colleges for critical thinking classes. And it's also why we have the meta-argumentative puzzle for straw man arguments—with straw man arguments, we reason about reasoning, we hold each other to scrutiny. But it's a fallacy. So how is it that all this reasoning about reasoning goes so wrong? It was supposed to have *good* results, like better first-order reasoning or clarity of how we evaluate where we are with each other in the dialogue. Instead, this reasoning about reasoning with straw man pulls us off the tracks. And so, we have a puzzle of meta-argumentation, and the straw man fallacy is a particular instance of the problem. Seriously, this is a secret. DO NOT tell the deans or the university donors.

To start, recall our point about reasoning as requiring contrasts. For a reason to play the role it does for us, we need a picture of what the range of alternatives is. What is controversial about what occasions the need for an argument is what then needs to be addressed by the argument. That's the big idea behind contrastivism about argument. This is a point, we believe, that is not merely a pragmatic point about arguments addressing contested options (which is most certainly appropriate, if not required as a matter of dialectical responsibility) but it is a point about the nature of reasons as reasons. For reasons to be *for*, there needs to be some alternative it sorts out as *against*. That's how reasons are contrastive at their core, and their quality is to be evaluated according to how effectively they make that contrast. A deductively valid argument can eliminate all contrasts with respect to what lies between the truth of the premises and the conclusion, so it needs dialectical support only with regard to the status of the premises (and, perhaps, to the assessment of it as valid). Inductive support requires that we ask what kinds of circumstances we face, and as new information comes in, the degree of support may increase or wane, because some alternatives may become, with that new information, more salient to address or less so. Contrastivism explains why that is the case and why context, as socially constituted features of our objectors and questioners and our ranges of intellectual responsibilities, requires we survey a range of relevant alternatives. And when we change who we are talking to, or whether they ask one question or other, a reason's probative quality can change. Dialecticality and context sensitivity is the social face of the contrastivity of reasons.

The consequence of this theoretical setup is that the evaluation of a reason's quality, in fact to see the reason as making any progress at all as a reason, requires we have a representation of the relevant alternatives. To know how good a reason is, we have to see its contrastive role. And to see that, we need to know what populates the contrast class. An upshot of this is that if one's intellectual opponents look less intelligent, with less impressive arguments, with manifestly objectionable views and easily answered objections, one's own reasoning begins to look very powerful. If reasons are contrastive and the contrasts are so easily refuted, then there's very little left but one's view as an easy winner. That is, the more the opposition to one's views appears as manifestly stupid, evil, or confused, the more confident one is tempted to be in one's views. Contrastivism explains how, then, negative misrepresentations of the opposition yields higher confidence in the quality of one's own reasons. Without this function, it's unclear why straw manning one's opponent seems to also be a form of praising and promoting one's own views. Now it seems clear, and now we can see why the straw man can have further downstream argument-closing functions.

All of this is a result of our reasoning about reasons. And if this theoretical explanation of the straw man is correct, the meta-reasoning here comes part-and-parcel with argument as such and with the straw man in particular. When we argue with each other, we share our reasons and jointly weigh them out. We are not just giving and asking for reasons, but we are evaluating them and their overall force in the process. And this requires that we reason about those reasons. Here's a simple way to put it: a central part of reasoning is reasoning about our reasons. And so, argument requires meta-argument.[4] This is why learning critical and evaluative terms like *irrelevant*, *strong*, and *rebutting* is part of reasoning together—we don't just pile all the reasons together and let it be some brute function of who's got more or just heavier reasons than the other, but a mapping out of the relations between them and the evaluation of their consequent strength. The *pro-con* back-and-forth of a properly run ongoing argument is arranged as it is precisely because weighing out reasons isn't a brute event, but one that requires multiple tracings of relevance and reply. When we reason, we reason about our reasons and their relations. What else would *counter-argument*, *objection*, or *reply* be? And, further, what else would *concession*, *revision*, and *reopening* be?

In order to evaluate reasons, we must represent them, and this is where we see the defining feature of straw man arguments—the straw man is made possible by the fact that as we reason together, we must reason about the reasons we've shared, and to do so, we must represent them. An incorrect representation of a dialectical opponent's contributions (as better or worse than they actually are) throws off the dialectical scorekeeping for the exchange. And though there can be salutary misrepresentations in both negative and positive cases, what's important is that we've identified where things *can* go wrong. We, again, do not hold that it is the *misrepresentation* with straw or iron manning that yields the fallaciousness of these arguments, but what we've called their *closing function*. Regardless of what makes the straw man fallacious, it is made possible because of this representational requirement of contrastive critical dialogue—if we exchange and weigh out reasons together, we must represent them in our joint evaluations to identify how they fit together as a whole. Bad arguments, in the process, get rejected, and good arguments are promoted and endorsed.

As we've noted, the straw man fallacy is exemplary of the class of *dialectical* fallacies. This general class is distinct from fallacies of meaning or relevance primarily because the fallaciousness of the argument does not lie simply in the premises, their relation to the conclusion, or to how they are interpreted. In fact, many of the arguments that comprise instances of the straw man fallacy may be perfectly acceptable when evaluated according to these criteria. Dialectical fallacies, instead, are fallacies because they are failures of argument-responsiveness with an interlocutor. Again, one needn't have a dialectical opponent to, say, assert the consequent or commit hasty generalizations. But in order to straw man, one must first have an opponent who's said some things or made some arguments. Straw manning occurs in light of these facts, and without there being a dialectical opponent about whose arguments one can give an argument, one cannot straw man. That's what makes straw manning a dialectical and meta-argumentative fallacy. The straw man is an *argument about someone's argument*. Now consider the following.

6.3.1 Carson's Kindergarten

How one can straw man someone's view or argument and then reason about those reasons can come in a variety of ways, but we will start with the *representational* straw man fallacy. Again, what one does is represent one's dialectical opponent's views in worse or less defensible form than what the opponent had given, and then one criticizes the representation of the opponent's views. Consider the way *Salon*'s Sophia Tesfaye represents Ben Carson's (at the time a candidate for the 2016 Republican nomination for president) claims about guns in schools. Carson, on ABC's *The View*, makes the following statement:

> If I had a little kid in kindergarten somewhere I would feel much more comfortable if I knew on that campus there was a police officer or somebody who was trained with a weapon […]. If the teacher was trained in the use of that weapon and had access to it, I would be much more comfortable if they had one than if they didn't […].

Tesfaye's headline to report what Carson said:

Carson wants kindergarten teachers to be armed.[5]

The force of the headline is that Carson is making a policy proposal and positively *wants*, as opposed to *prefers*, armed kindergarten teachers. Moreover, there is a difference between preferring that if a teacher had firearms training that they are armed at school and arming kindergarten teachers in general. One may still reasonably disagree with Carson on the matter, but the disagreement needn't now be with someone so extreme.

What the straw man does in this case is not only misrepresent Carson's case for armed teachers (in having a preference for a teacher already with firearms training to bring a gun than not) but it misrepresents Carson's track record and prospects for improved dialogue on the matter. It closes the issue, because we reason that the view and those promoting it have little hope for improvement. The view is clearly wrong-headed, and those who hold it are, too. Tesfaye concludes that "Ben Carson is just this vile." The meta-argumentative move, given the evidence of the negative misrepresentation of the straw man's target view and speaker, is to make the argument that it's time to close up shop on continued discussion on this matter—if this is as good as it gets with the other side, the debate is effectively over.

We've termed this move with straw man arguments their *closing function*, and they work the way they do because we are reasoning about reasoning and reasoners. The meta-argumentative fronts for the straw man are *retrospective* and *prospective*. On the one hand, the straw man speaker misrepresents the target's argument and evaluates the argument as bad. This is the *retrospective* meta-argumentative face of the straw man. It establishes a bad track record to reflect on when we assess arguments (and their prospects) going forward. On the other hand, the speaker then makes the case that, given the standing evidence about the state of dialectical play (or what the dialectical score is, overall), continued argument on this subject with these targets is hopeless. So the discussion is closed. We've shown that it's *here, in inappropriate closure of argument*, that the straw man is fallacious, as it is possible for the retrospective meta-argumentative misrepresentations to spur and open new avenues for discussion so long as we do not make the prospective meta-argumentative closure. And it all is possible only under the condition that we are reasoning about our reasoning—what reasons we've given and what reasons we expect to give (and receive) going forward.

Consider the meta-argumentative valence of what we've called the *hollow man*. Again, one does not take any particular opponent's view and distort it but rather one just invents a ridiculous view for one's opponents whole cloth.

6.3.2 Prager on Gun Control

In October of 2015, *National Review* columnist Dennis Prager portrayed "The Left" in the ongoing debate about gun control in the United States as not being particularly reflective. According to Prager, it comes down to three things behind their commitment:

> The Left believes in relying on the state as much as possible [...]. The Left is uncomfortable with blaming bad people for bad actions [...]. The Left is more likely to ask "Does it feel good?"[6]

Who Prager has in mind in this case is not clear. Not once does he reference anyone who has social-scientific data or who points to the fact that there are fewer gun-related crimes or large-scale killings in countries with extensive gun control. So he's not paying attention to the leading arguments in the domain. Instead, for Prager, the gun control debate really comes to having to answer people who want government-dependent, irresponsible, hedonists for citizens. Of course, the debate won't look very good when your opponents are so depraved.

Notice that straw manning requires a retrospective form of misrepresentation of the overall intellectual situation in an area of dispute. The speaker, the one who straw mans, must portray the target opposition in an untoward light. And in this case, it is a matter of attributing to them reasons that are manifestly bad. Not only are the reasons bad, but they tell against the moral and intellectual character of those who hold them. Consequently, a meta-argument goes that not only are those retrospective reasons bad but because of this, we should expect little good from this group on the issue. For Prager's audience, the third-person version of the straw man closes the discussion with the prospective meta-argumentative inference. This requires, as we'd noted earlier, an audience that either must not know better or is exercising the hermeneutics antipathy along with Prager. That is, if the audience for a straw man argument knows that and cares about the fact that there are better versions of the view available, then the argument will simply not work on them. And so straw man arguments of this form depend on their audience's being generally unfamiliar with the issue or at least with the opposing view on the issue. And so, with the straw man argument, they are given enough evidence about the debate (and, in particular, one side of the debate) to decide the matter. Again, this was our solution to the effectiveness puzzle with straw man arguments—that in the third-personal form, they must key either on the ignorance or antipathy of the audience.

This fact, that straw manning of this form depends on audience ignorance or antipathy, is significant. This is because, with straw manning, not only is a conclusion about the opposition and overall state of dialectical play on the issue established fallaciously, a picture of the opposition is painted in a way that yields and contributes to intellectual contempt. A result of this picture of the opposition as incompetent or mendacious is that one is less likely to want to engage honestly with them in further discussion. And as a consequence, we see the polarization of discourse on matters of significance. Straw man arguments not only produce bad argumentative results at the times they are given but they have lasting repercussions on the communities they convince.

Now a curious fact about this polarization of straw manning is that it runs in both directions. Prager constructs a hollow man view of "The Left" (if you don't like this example, just replace "Left" with "Right"—there are loads of examples) that few or any on The Left recognize or affirm. Naturally, they're not the target audience but they are *an* audience. The target audience, the dupes who believe this, conclude that there is no point in conversation with The Left. The net result is that they don't encounter The Left's actual views. Further, in its turn, The Left, or whatever parts of it that hear this argument, do not show up to contest it on account of the straw manning. That is, they know that if they *do* contest it, they will face versions of it in the second person—which is argumentative heavy weather nobody really relishes. So they sit it out—and so there is silence on the issue from the targets. The straw manners and their audience then take the silence of the target as their acquiescence to the argument. And so, curiously, the third-person straw man has an interesting looping effect back through the second person. Their absence from what would be a pointless debate is seen then as further evidence of the truth of the initial claims. Again, because the targets don't defend themselves, it's taken to be further meta-evidence of the correctness not only of the picture painted of them but of the correctness of the views of the straw manning speakers on the issue. (We will speak more explicitly to the meta-argumentative element to this line of reasoning shortly.)

Our case is that straw man arguments have the meta-argumentative valence they do in the third person, since they provide the audience with evidence of the quality of the target's views, reasons, and by extension, their intellectual character, as we noted in Chapter 3. One reasons retrospectively about the target on these matters and infers that the prospects for the view or critical discussion to be improved are not good. And so the speaker and audience move on and leave the target in the argumentative dust heap. The straw man does this, again, because it is an argument about arguments and arguers with two faces: retrospective, about arguments *given*, and prospective, about arguments *to come* (if at all).

This is seen, too, in second-person forms. In the dialectically silencing second personal form, the speaker straw mans the target to the target, but for the sake of making the argument too costly for the target to maintain. And so the straw man in this instance provides the retrospective argument that the speaker has interpreted the target's contribution uncharitably and prospectively promises that more uncharitable uptake is coming. The argument is going to be more like combat. And the same goes for second-personal gaslighting straw man arguments. The speaker provides a manifestly negative representation of the targets' contributions to the targets so that, retrospectively the targets will see themselves as having not exercised good judgment in the dialogue, and so prospectively infer that they are not good participants for further critical engagement. Again, straw man arguments can do that only if they are arguments about arguments, and by extension, the arguers who give them. The meta-argumentative result, then, is that, with each one of these cases, the straw man representations work as track records licensing the *closing function* of straw men argument. The problem is that this is a hasty and unjustified foreclosure of continued exchange on the issue. And it is this feature of the straw man that we think is the fallacious element.

6.4 Meta-argumentation and Meta-argumentative Fallacies

Logic is a meta-language to the first-order natural language in which we usually reason. But even in that natural language, we have signals that identify and reflexively endorse the argumentative moves made. Again, vocabulary like *so* and *therefore* don't just announce inferences, but they also endorse them as legitimate. They are normatively gravid, or given that Latin terms make all things in logic better, they are *igiturgid* (Latin's "therefore," *igitur* + *turgid*, anybody?). With the development of the meta-languages of argument evaluation, we have terms like *premise, conclusion, undercutting defeater,* and so on. And we have the concept of *fallacy*. The idea of developing and deploying the idea of *fallacy* is a good one, and we'd given an extensive defense of the practice of not only theorizing fallacies but of expanding the canon of them in Chapter 1. In short, it sharpens our interpretive vocabulary, and the list of fallacies is a heuristic tool for improving critical discussions with ways of testing argument quality. This is all a path to sweetness and light for rational exchange. There are, however, temptations to the dark side.

One temptation that comes with first having learned logic, and the list of fallacies in particular, is familiar to most any sophomore in our critical thinking classes. It is to commit what we've called earlier the *Harry Potter Fallacy*. Again, Harry Potter and his witch and wizard friends of J. K. Rowling's universe are given magical powers of self-defense by being taught a few Latin phrases and learning to flick their wands just so: *Expelliarmus!* Well, many students think that the fallacy list works the same way—they learn Latin terms like *ad hominem, ad verecundiam*, and *ad baculum*, and when necessary, they whip them out and by pronouncing them properly, disarm and petrify their intellectual opponents. The problem is that these terms don't work like magic words. Our point is that the Harry Potter Fallacy is a unique fallacy of meta-argumentation. Once we learn the vocabulary of argument evaluation, new errors are made possible, because now we not only can err when we think about things, we now can err when we think about how we think about things! So, there are not just *argumentative fallacies* that arise in argument but there are *meta-argumentative fallacies* that arise when we argue about argument. Here is the temptation: it is a plausible line of thought that labeling an argument as a fallacy is enough to criticize it. So, it works like a magic word, or at least like an endorsement from a judge with standing. That's why all the Latin matters—it's got *gravitas*. But as it turns out just invoking the name of the fallacy one thinks one sees in an opponent's argument does not actually provide reason for rejecting it. Even in Latin. The fallacy name is a placeholder for the work of explaining what is wrong with a premise or the failure of support the premises provide. That's the Harry Potter fallacy. Shouting *"ad verecundiam!"* is as effective as shouting *"Expelliarmus"* for muggles in a disagreement. That is, not very. Now, that the Harry Potter Fallacy is a fallacy is itself interesting, but what's more interesting for our purposes here is that it shows that learning the meta-language of logic *can* improve our reasoning, *but it can also occasion new meta-argumentative fallacies*, fallacies we commit when we reason about reasoning. Now, *that* is interesting! (But it's also a little depressing, and again, it's a secret that must be kept from the deans and philosophy department donors.)

The straw man is, as we've argued, a meta-argumentative fallacy, and it's one that arguers can commit without too much meta-linguistic overhead. All one needs to have is the rough idea of worse and better arguments, and that's about it. But what happens when one combines the conditions that yield the Harry Potter fallacy with the phenomenon of the straw man? That is, what might be the case if the straw man becomes a meta-meta-argumentative fallacy—in particular, one we commit when we use the meta-language of logic to argue about not only our arguments but our arguments about arguments? The qualifier *meta-meta-argumentative* is a mouthful, and maybe a little overboard, so let's call it a *strongly* meta-argumentative fallacy. To be a *strongly meta-argumentative fallacy*, as we see it, the argument must use the meta-language of logical evaluation as an essential part of the meta-argumentative fallacy. You can't commit this kind of strongly meta-argumentative fallacy without first learning the meta-language of logic and then misusing it in arguing about arguments. So, with the strongly meta-argumentative straw man, the speaker's misrepresentation of the target's view and premature closure of the exchange depends on the speaker explicitly casting the target as committing a fallacy—they go so far as to name the fallacy, and then they move on with the discussion as though the matter is closed. That is, learning the fallacy list becomes a menu for straw manning one's opponents.

6.4.1 Ann Coulter's *Ad Hominem* Charge

One way that the fallacy list allows for what one might call a menu of straw men is that it gives uncharitable interpreters a set of forms to aim for in their negative representations. If you can interpret the opposition to just be speaking in identifiable fallacies, then the work of critique will be easier and the closure of the debate will be more decisive. The American conservative political commentator Ann Coulter's *Slander*'s thesis is that "*ad hominem* attack is the liberal's idea of political debate" (2002: 13). She describes the regular exchange with liberals as follows:

> The "you're stupid" riposte is part of the larger liberal tactic of refusing to engage ideas. Sometimes they evaporate in the middle of an argument and you're left standing alone, arguing with yourself. More often, liberals withdraw figuratively by responding with ludicrous and irrelevant personal attacks. Especially popular are *non sequiturs* that are also savagely cruel. A vicious personal attack, they believe, constitutes clever counterargument.
>
> (2002: 153)

The problem, as Coulter sees it, is that liberals keep calling conservative views, proposals, and conservatives themselves "Racist!" "Sexist!" and "Homophobic!," and this is as good as the challenges get. Again, as we noted in Chapter 3, Coulter nutpicks weak men for cases such as this. So, though many liberals may be insult machines for conservatives, it doesn't follow that the overall case liberals make against conservatives is one of *ad hominem* abuse. That Coulter has weak manned the opposition should be clear, and that's interesting as a different form of straw man. That's a point from a previous chapter. The issue here is that the *tools of critical thinking* were the means that allowed Coulter to make this case. That is what's the object of our attention now. So, there's a fallacy (in this case, a weak man), but then the fallacy is committed with the vocabulary of identifying fallacies. The weak man that Coulter selected was one made out to be a collection of *ad hominem* attacks by her targets for criticism. It was because Coulter knew about the fallacy of *ad hominem* abusive as a fallacy, and because her audience knew about it, too, that this weak man depiction of liberals and their arguments is as powerful as it is. And notice not just how Coulter uses the meta-vocabulary of argument-analysis (*ad hominem*, *non sequitur*, and the Millian norms attempting to continue discussion) but she shows that continued argument yields little improvement. "[Y]ou're left standing alone arguing with yourself." This particular weak man argument is, as we've termed it, a *strongly meta-argumentative fallacy* of straw man (or more particularly, the weak man). The fallacy is what it is and has the presentation it does precisely because the speaker and audience are aware of and use the categories of meta-argument—in particular, the vocabulary of fallacies. The tools of critical thinking just made a new kind of fallacy possible. That's amazing and alarming, right?

So, given that straw man is a meta-argumentative fallacy, and meta-argumentative fallacies can be strongly meta-argumentative when they are posed within the categories of the meta-language of meta-argument (especially with the fallacy terms), we have a big question: *Does the concept of the straw man itself play a role in strongly meta-argumentative straw manning?* Our answer is: *Of course it does!* All that's necessary is for the accusation of *ad hominem* or other fallacy charge in the representation (as we saw earlier) be replaced with accusations of straw man. Consider the following.

6.4.2 Obama the Demagogue

Victor Davis Hanson, over at *The National Review*, is an accomplished classicist. Hanson's July 11, 2011 column, "The Demagogic Style," was a short account of the early usage of "demagogue" and "demagoguery" from Thucydides through Xenophon and Aristotle.[7] Once the diagnostic and evaluative apparatus is in place, Hanson turns to look at how the demagogic style works in President Obama's rhetoric. One tactic that caught our eye was the strategic use of charges of straw men, in particular, Hanson's third on his list of demagogic moves:

3) The evocation of anonymous straw men, sometimes referred to as "some" or "they."

In the Manichean world of Barack Obama there are all sorts of such demons, mostly unnamed, who insist on extremist politics—while the President soberly and judiciously splits the difference between these fantasy poles. So for the last three years we have heard, but been offered few details, about the perils of both neo-con interventionists and reactionary isolationists, of both profligate big spenders and throw-grandma-over-the-cliff misers, of both socialist single-payer advocates and heartless laissez-faire insurers who shut emergency-room doors to the indigent in extremis—always with the wise Barack Obama plopping down in the middle, trying, for the sake of all the people, to hold onto the golden mean between these artificially constructed zealots.

Hanson has provided an interesting analysis of how demagogic "moderates" sell their ideas—they portray themselves as avoiding the vices of two caricatured extremes. The trouble, as Hanson sees it, is that nobody actually occupies those extremes that Obama situates himself between. These extremes that are to be split are men of straw. In fact, in our lingo, Hanson's charge is that Obama is *hollow manning,* since he says *nobody* has a view anything like what Obama is invoking. But a few things. First, so far, all Hanson has done is say that the positions are anonymous in Obama's presentation. That doesn't mean that they don't have occupants. That just means that the president doesn't always have to name his dialectical opponents. That's an old rhetorical advantage presidents have always had—they are presidents and so naming and tangling with reporter X or talking head Y would be a problem for the office as an executive governmental position. For what it's worth, the presidency is a democratic approximation of regal leadership, and so sometimes occupants of the office don't have to name their targets for critique. Knowledgeable interpreters can fill in the blanks. Second, Hanson is way off in the interpretation if he infers that given that since President Obama doesn't name names, it means that nobody actually occupies that position. It was quite easy in 2011 for us (or most any politically informed reader, for that matter) to give names of prominent spokespeople for these extreme sides Obama was trying to find a middle between. For example, former Arkansas governor Mike Huckabee was the leading representative of the "personal responsibility" right on medical coverage, and Vermont senator Bernie Sanders for the "single payer" left. Done, and no problem at all, really. It just takes some familiarity with the terrain, and we can easily populate the rest of those extremes for ourselves. Reading a newspaper or watching thirty minutes of cable news talking heads should do it.

Our own view is that the hollow man trouble with Hanson's representation of Obama's setup isn't really that the extremes are anonymous or that they aren't populated but that there is a lot of ground to occupy between the extremes. And when one sets the extremes up in that way, anyone can look like a moderate. That charge would have been a more appropriate charge against Obama at the time, but that would not have fit the mold for *Hanson's own straw man of Obama's position*. And so, Hanson has straw manned Obama by saying he has straw manned (again, more accurately, hollow manned) his opposition. Hanson not only commits a meta-argumentative fallacy, one that is a fallacy one commits when reasoning about reasoning, but he commits this fallacy using the meta-language of articulating the norms of reasoning. Thus, it is not only a meta-argumentative fallacy but a *strongly meta-argumentative fallacy*. You just can't commit this kind of argumentative error Hanson has committed it unless you've had the requisite training of both naming and explicitly avoiding committing these kinds of errors. It's an error of reasoning only the most learned of us can commit. It's all very ironic, and a little impressive, really.

So far, we've seen straw man meta-arguments be about how the targets of the straw man argument themselves straw man *others*, but we can also straw man a target in terms of how they straw man our own view. One way to do this is to portray the opposition to be deeply confused about what your view at issue is and what the relevant considerations around it are. And to do this, the speaker offers himself up as the straw man target for his own straw man target's argument. It's all very complicated, because it's all very meta. So an example will help.

6.4.3 Goldberg's Strict Constructionists

In August of 2010, conservative Columnist Jonah Goldberg published an article at *Town Hall*[8] defending South Carolina senator Lindsey Graham's proposal for a Constitutional Amendment that would revise the Fourteenth Amendment's citizenship clause so that children born of undocumented immigrants are not citizens.[9] The details of the Amendment or its revision don't matter for our purposes, as the main issue of the argument was that Goldberg was out to defend our responsibility to revise and interpret the Constitution as the cases demand. Now, this should come as a surprise to all the conservatives who take themselves to be strict "Constructionists"—this sounds all too much like the old "living document" take on the Constitution that conservatives hate so much, associated with the liberals, who think that as a living document, the Constitution also guarantees more rights than explicitly named in it (most importantly, that of privacy). Goldberg anticipates this challenge:

> this "living document" argument is a straw man. Of course justices must read the document in the context of an ever-changing world. What else could they do? Ask plaintiffs to wear period garb, talk in 18th-century lingo and only bring cases involving paper money and runaway slaves?

Goldberg's a little confused about straw men, as straw manning depends on how you portray your opposition, not how obvious your own views are. As we take it, the obviousness of Goldberg's view is supposed to be implicated by the fact that he holds that the only objection to it could be on the basis of presentational issues of period dress and mannerism, which itself is a straw man of strict Constructionism. But the objections to strict Constructionism aren't on the basis of what one wears when in court but to the fact that identifying original and literal intent of normative language requires that one have clear ideas of what "happiness" and "freedom" meant to those drafting the Constitution. The problem is that they themselves disagreed over those matters. So Goldberg's rhetorical flourish here is just high-minded jibber-jabber, and clearly a straw man of those who criticize strict Constructionism. Though Goldberg's attributed straw man of strict Constructionism is clearly a straw man itself of what critics of the view would say in their objections, the key is to focus on the question of who would think that it's a matter of one's clothing and locution to interpret the Constitution. That, of course, is the key joke on Goldberg's part, as straw men are supposed to be so ridiculous and absurd, one can't help but giggle at the very idea of someone believing that nonsense—and so, one gets a cheap laugh at the target. In this case, we object. Not just because we care about well-run argument but because we care about well-run comedy. Regardless, Goldberg's overall point is reasonable enough—if the options are, on the one hand, seeing the world and the Constitution's relevance through the lenses of eighteenth-century Yankees and, on the other hand, looking at the world with the judgment of twenty-first-century Yankees, we should take the twenty-first-century perspective, given that we're out to deal with twenty-first-century problems. So Jonah Goldberg has made a nice point about interpretation and also has highlighted a straw man argument of how strict Constructionism could be erroneous. The joke in the service of the point isn't so good, but we can let the bad comedy go. This is because the strongly meta-argumentative fallacy of straw manning with the straw man accusation is Goldberg's primary move:

> When discussing the Constitution on college campuses, students and even professors will object that without a "living constitution," blacks would still be slaves and women wouldn't be allowed to vote. Nonsense. Those indispensable changes to the Constitution came not from judges reading new rights into the document but from Americans lawfully amending it.

Even professors would think that a living Constitution is necessary for these results? Surely only professors not familiar with the Constitution *at all* would say that sort of thing, because the Constitution was amended to officially end slavery and extend suffrage to women (with the Thirteenth and Eighteenth Amendments, respectively). And those are pretty famous Amendments. Goldberg owes us at least one name for this charge, since it's not clear that anyone keeping up with the discussion could call one to mind easily. But he provides no documentation, no names, no nothing, just vague allegations of intellectual incompetence. Nobody said that the living document interpretation of the Constitution was the solution to *those* things—we had Constitutional Amendments to solve those problems. Only dunces would say those were cases of living document work. But how about, say, *Brown v. Board* over integrating schools, or pretty much every privacy rights case? Or, maybe *Gregg v. Georgia*, with the notion of an evolving standards of decency in punishment? Those are all cases of reading the document of the Constitution in a way that keeps its core commitments but also extends them to the cases that the framers did not anticipate. Ignoring these cases (and actual discussions of them on academic campuses) not only distorts what the "living document" interpretation is but it makes it impossible to make sense of what Goldberg's own views on the Constitution are, since the only other alternative given his setup is strict Constructionism. For someone out to prevent straw manning about Constitutional interpretation, Jonah Goldberg is an expert at constructing and knocking the stuffing out of straw men he constructs. And this, given the importance of contrasts to identifying reason quality, makes it hard to figure out just what Goldberg's reasons amount to, since in the setup, there's a straw man of his view attributed to a set of critics, and then, thereby, a straw man of the critics with that representation. A careful reader should emerge from the exchange with the question of whether there was any positive view presented or defended, since it was all simply debunking comically bad views.

Our point is that Goldberg constructs a straw man of his opposition by putting a straw man of his own position in their mouths and then calling them out for straw manning him. This, again, is all very meta-, but that's what happens with reflective and fallible beings who argue about arguments. And this is not just a fallacy but it's what we've called a *strongly meta-argumentative* fallacy. Learning about the straw man fallacy is what creates the possibility of new kinds of straw men of one's opposition—that they commit the straw man fallacy when they criticize Goldberg and his allies with the arguments they use. Of course, as we've noted many times, this kind of strategy is generally a *third-personal* form. Goldberg invokes *professors* who move forward with this bad argument against him and misunderstanding of his view, but it is clear he is not out to address them as they see their own views or to correct them. Rather, he is addressing his preferred audience at *Town Hall*, already keen to see their political opponents portrayed as intellectually vicious. And professors are one of their favorite targets for contempt. And in this case, the opponents are portrayed as not only ignorant of the Constitution and US history but they are straw manners of those who are right about so many things.

So the key to this strategy of strongly meta-argumentative straw manning with the concept of straw man is for a speaker to convince an audience that the targets of criticism (of the speaker's own straw man) have themselves straw manned the speaker. The move from this is to infer that if a straw man of this (the speaker's) position is the best objection to the position going, then critical opposition to this view is insubstantial. Assuming there is some reason supporting it, the speaker's view should deserve some credence. And, for sure, the opposition to it is undercut. That's the closing function of strongly meta-argumentative straw man arguments, and it is important to note that so far, we've shown that this requires that most of the targets for these straw men are not given a good deal of exposure. The audience needs only enough to see the objectionable misrepresentation, and no more.

But now consider, by analogy, *the dive* in soccer. Soccer players, if fouled in the opposing team's penalty box, are given a penalty kick—an unobstructed shot at the opponent's goal from twelve yards out and with only the goalkeeper in the goal. It is a very valuable scoring opportunity. And so, it is a regular phenomenon in soccer games to see players dribble the ball into the penalty box and then *try to get fouled*, or worse, they initiate contact with a defender and then *act as though they've been fouled*. That is a *dive*. And if the referee is a convinced audience for all this soccer drama, a penalty kick is awarded.

Argumentative diving is a form of strongly meta-argumentative straw manning. It requires that the speaker convince the audience that the target has straw manned *him*, and so exercises inappropriately uncharitable interpretive attitudes, and so is not worth further inclusion in properly run dialogue. So what the speaker will do is *bait the target with an argument that will incite them*. The target will react negatively to the argument or claim, and then the speaker will turn to the audience and invite them to note how unfairly he is being treated. And so the target for straw manning is charged with straw manning, but they have only responded to an objectionable claim from the speaker. It's a dialectical trap for the target and a piece of argumentative drama for the audience. So it is what we would call the *self straw man*. Consider the following instance.

6.4.4 Stormy Daniels

President Trump has been the target of a defamation lawsuit by Stormy Daniels, an adult film star who allegedly had an affair with Trump years ago. Because Trump consistently publicly insulted Daniels and her attorney denying the affair, Daniels had pursued legal remedies. Daniels' lawsuit was dismissed by a judge, and thereupon Trump went to Twitter:

> "Federal Judge throws out Stormy Danials (sic) lawsuit versus Trump. Trump is entitled to full legal fees." @FoxNews Great, now I can go after Horseface and her 3rd rate lawyer in the Great State of Texas. She will confirm the letter she signed! She knows nothing about me, a total con![10]

The key to our analysis here is that Trump calls Stormy Daniels "Horseface" when announcing the case is dropped. The president has a long history of saying nasty things about women's appearances, so he was asked about it by the Associated Press in an interview shortly following the tweet. Trump did not back down from derisively nicknaming Stormy Daniels "Horseface" hours earlier, but he invites the reporters interviewing him to interpret the tweet themselves.

> He says "you can take it any way you want," when asked if it was appropriate to insult a woman's appearance.[11]

The question wasn't what the statement *meant* but whether the president *stands by* the statement given what it clearly means. Ironically, the reporters allowed Trump to *retract* the statement. It was clearly an insult. The most charitable interpretation was, at the time, that the tweet was out of anger or elation. One can retract those statements or put them into context of simple jubilancy, but one can then take it back. Because it was an unequivocal insult. Trump's invitation to "take it in any way you want" implicates that there are other ways to interpret the statement. But what are the options for our preferences to interpret this statement, to begin with? Is there another option, perhaps less misogynistic, to interpreting calling a woman "Horseface" to be a way of maligning her looks? Maybe it's a shorthand that rich guys use to show that they know someone who looks like they own horses... you say, "Ah, Sterling... he clearly has a wonderful set of stallions at home... *see his regal horseface*?" But it is still hard to take it in these (iron manned?) lights when the expression is immediately before calling Daniels' attorney, Michael Avenatti, a "third rate lawyer". It's an insult both by itself and in the parallel construction with another insult. And it is in the context of a tweet of celebrating one's victory and taunting one's political rivals—surely this is the time when a man who denies that he had an affair with a woman would overtly malign her. It's not just clearly an insult but, again, one about a woman's appearance. Competent interpreters of these kinds of communications can see what's happening very easily.

What is important, though, is that Trump's response inviting reporters and other readers to "take it any way you want" is a trap. The trap is as follows—if Trump has said that we can interpret the claim as we see fit, then if we interpret it as offensive, that's evidence that we've *chosen* to interpret the claim as something objectionable. But who would do such a thing, except someone who suffers from an irrational, uncivil bias? And so, by saying that this unqualifiedly objectionable piece of language can be taken as we wish, Trump, by his lights, is testing us for whether we choose to blindly resist him on everything and act deeply offended when we do just that, or we simply see that Stormy Daniels is as ugly as he says she is, and we agree. But the point, again, is the trap: once you *choose* to be offended by interpreting his statement in the offensive way, how is he *really* responsible for the objectionable stuff? The only apology he would owe, then, would be that he's sorry that people can't help themselves but to interpret him in a nasty way all the time. And so Trump may turn to his base and point to the "Trump Derangement Syndrome" of those who criticize him for misogynistic tweets. This is, really, our point again, with the hermeneutics of antipathy, but now even this point is weaponized as an observation about how this interpretive stance yields straw men. In fact, it was precisely how Trump was defended by Jesse Watters on Fox News, arguing that "the Left" is "clutching their pearls when the president uses the word 'horseface.' Please. To be surprised now that Donald Trump says bad things and uses dirty nicknames about people three years later means you either haven't learned a thing or you're just dense."[12]

With charges of straw man, those who make the challenge take on particular dialectical burdens. One of them is to point out how the view that has been straw manned is not only better than the representation but that the better view was accessible to those who performed the straw man. Namely, that a reasonable interpretation was available that did not suffer from the problems with the represented view. But here's the problem with the Trump case here with the trap: he hasn't offered any alternative that's a reasonable interpretation that's not misogynistic.

A further strongly meta-argumentative use of straw man is in terms of characterizing the opposition as indistinguishable from the straw men one is regularly tempted to make of them. Poe's Law, an eponymous law of the internet, applies to cases where one can't distinguish satire of a view from a real view with an explicit sign that it is satire. So, for example, liberals, before the 2008 presidential election, had difficulty telling the difference between parodies of Sarah Palin, the Republican vice presidential candidate, and direct quotes. And conservatives consistently could not identify comedian Stephen Colbert's *Daily Show* segments as parodies of conservative viewpoints.[13] That their own views are indistinguishable from parodies of them (even by them) is taken as telling. That is, damning. Once we see that straw men of views can then be served back as versions of sincere and best statements of these commitments to those who hold them, a new meta-argumentative move is made possible. It takes the following form of experiment: Can those who hold a view distinguish their view from a straw man of their own view? And thus initiated what might be called *the hoax literature*.

6.4.5 The Hoaxes

The pomp and circumstance of academia notwithstanding, scholars regularly accuse each other of having not just false but completely ridiculous, unsatirizable views. Alan Sokal, a physicist at New York University, thought as much of the "postmodernism" in cultural studies and he concocted a plan to reveal it by submitting an article filled to the brim with postmodern nonsense to *Social Text*, a leading journal in the field. The thought was that, should the article get accepted (which it did), even representatives of the best judgment of the field can't distinguish mockeries of their views from sincere expressions. The essay he wrote, "Transgressing the Boundaries: Towards a Transformative Hermeneutics of Quantum Gravity," argued that theses in semiotics and speculative psychoanalysis were confirmed by discoveries in quantum field theory, and then offered the idea that the axiom of equality in set theory was justification for radical social democratic political conclusions. These are errors anyone who knows anything about these issues would see as equivocations at best. The overall thesis of the essay was that gravity was a social fiction, and this discovery undermined the very idea of existence. What's left, Sokal argued in the paper, was "liberatory science," freed of demands for evidence, truth, or even critical review. The essay was clearly an instance of irresponsible "theory."

The editors at *Social Text* accepted the article. Sokal then went to *Lingua Franca* to tell the academic world about it.[14] In *Social Text*'s defense, they didn't practice peer review at the time, so the bar was extremely low. That said, this defense isn't much of a defense, as not peer reviewing is itself a self-damning explanation. It would be as though saying the shepherd who guards the sheep has a good reply to those who say he's irresponsible for letting foxes take two ewes when he says he wasn't paying attention at the time. But we digress. The lesson taken by many from what's been called by many *The Sokal Hoax* was a meta-argumentative one: the postmoderns, in seeing a straw man of their own views as a legitimate contribution to their ongoing research program reveal that they're really just fans of fashionable nonsense. Again, the key is that Sokal did not give the meta-argument *to the targeted group*, but he published the result of his hoax elsewhere. It was a meta-argument with an audience already suspicious of the postmoderns. And Sokal went on to publish a book about it all with Jean Bricmont, under the title *Fashionable Nonsense* (1997).

Inspired by Sokal's earlier hoax, in 2018 Peter Boghossian (of Portland State University's Philosophy Department), James Lindsay, and Helen Pluckrose (an editor of *Areo*, an online journal) aimed to get Poe's-Law-style pieces through some semblance of peer review.[15] Under various pseudonyms, the trio wrote and submitted a series of articles to several journals in sociology, cultural studies, and (we're sorry to report) philosophy. In all, they wrote twenty papers and managed to get seven accepted (one of them by *Hypatia*, a well-known feminist philosophy journal) and four actually published with additional papers receiving revise and resubmit judgments (before the hoax was uncovered). The three published their findings at *Aero* with the title, "Academic Grievance Studies and the Corruption of Scholarship." They framed their project as targeting work in a domain they had termed broadly "grievance studies," which they represent for their readership as:

> fields of scholarship loosely known as "cultural studies" or "identity studies" (for example, gender studies) or "critical theory" because it is rooted in that postmodern brand of "theory" which arose in the late sixties. As a result of this work, we have come to call these fields "grievance studies" in shorthand because of their common goal of problematizing aspects of culture in minute detail in order to attempt diagnoses of power imbalances and oppression rooted in identity.
>
> (2018)

Their plan was to show that "our papers are all outlandish or intentionally broken in significant ways, it is important to recognize that they blend in almost perfectly with others in the disciplines under our consideration." If the papers are accepted at what they assessed as "significant and influential journals," then they have shown that the work that is produced in these fields, "*is not* knowledge production; it's *sophistry*. That is, it's a forgery of knowledge that should not be mistaken for the real thing. The biggest difference between us and the scholarship we are studying by emulation is that we know we made things up." And so they proceeded to write what would first "use what the existing literature offered to get some little bit of lunacy or depravity to be acceptable at the highest levels of intellectual respectability within the field," and then turn to argue for wild and unacceptable theses. The titles and theses ranged along the following lines:

> "Human Reactions to Rape Culture and Queer Performativity in Urban Dog Parks in Portland, Oregon": That dog parks are rape-condoning spaces and a place of rampant canine rape culture and systemic oppression against "the oppressed dog" through which human attitudes to both problems can be measured. This provides insight into training men out of the sexual violence and bigotry to which they are prone.
>
> "Moon Meetings and the Meaning of Sisterhood: A Poetic Portrayal of Lived Feminist Spirituality": No clear thesis. A rambling poetic monologue of a bitter, divorced feminist, much of which was produced by a teenage angst poetry generator before being edited into something slightly more "realistic" which is then interspersed with self-indulgent autoethnographical reflections on female sexuality and spirituality written entirely in slightly under six hours.
>
> (2018)

The last one was written "to see if journals will accept rambling nonsense if it is sufficiently pro-woman, implicitly anti-male, and thoroughly anti-reason for the purpose of foregrounding alternative, female ways of knowing." It got accepted. In their conclusion, Boghossian et al. ask what they have demonstrated, and they invite their readers "to make up your own minds about that." Of course, given their readership (since they did not submit their report on the hoax to one of their target journals, but to the readership at *Aero*), it should be clear what that conclusion would be. They quickly follow up, noting: "it shows that there are excellent reasons to doubt the rigor of some of the scholarship within the fields of identity studies that we have called 'grievance studies.'"

There are lots of kinds of hoaxes, and all of them rely to some extent on misrepresentation. But the central feature of this sort of hoax is strongly meta-argumentative. In fact, Boghossian et al. remark how this is a kind of meta-argument that requires that the targets for evaluation are part of the evaluation, as the authors hold that they are performing a form of "reflexive ethnography" of this field and those working in it. The point isn't merely to describe a view as silly but also to show that people who share the target view are not up to the task of sorting out good arguments from bad ones within the domain they purport to have expertise. And so their acceptance of an obvious straw man of their own views shows this unhappy truth not only about them but about their field of training.

Oftentimes, the success of hoaxes of this representational sort depends on social factors, such as politeness, or clever editing of reactions. To some extent this was true of the Grievance Studies hoax—some reviewers, exercising abundant charity to the hoax ccontributions, took them to be in earnest, so they provided ample feedback. Nonetheless, it's worth stressing a couple of points about these Sokal-type hoaxes. First, journal reviewers must have ample time to consider and reflect on what is being presented to them for publication. There is even an incentive in reviewing to afford heightened scrutiny for things submitted for publication. Second, the hoaxed view is not asking merely to be taken seriously by a listener on the spot but rather to be published as representative of a meaningful contribution to the relevant discourse. The idea, of course, is not, as one of the hoax authors suggested, that the articles are in some sense true or that their publication stands as a kind of ringing endorsement by all those in the field; this isn't what journal articles are about in these fields. Rather, the idea is that they pass a kind of muster such that they're worthy contributions to ongoing debates. That is a considerably lower bar.

The straw man normally, and infuriatingly, works by presenting the distorted version of an argument to a willing or perhaps unprepared audience. With no one able or willing to reply, they have quick uptake by their audience. Normally, as we've seen with the effectiveness puzzle, such manipulations won't work on their third-personal targets. And the main reason why is that *the straw man wasn't for them*. Indeed, in the present case of the grievance studies hoax, the hoax itself worked on the *National Review*, a conservative publication with a "campus watch" section editor, who took the author of one of the academic hoax papers seriously, writing:

> A paper written by Portland Ungendering Research Initiative's Helen Wilson claims that dog parks are actually very sexual places where we can learn things about rape culture and "queer performativity.[16]
>
> Yes—seriously.

Well, not actually seriously. But it's not surprising or revelatory that the *National Review*, and particularly with its section editor being given "campus watch" duties, would find it ridiculous. It's the fact that the hoax circumvents the effectiveness puzzle by inducing the target not only to agree but to participate in the critique of their own position and their own standing as arguers. To make this all the more horrible for all involved, the paper just mentioned on canine rape culture in dog parks even won an award.

It bears repeating that the point is not that the views of the targets of the hoaxes we've surveyed are bad or that their defenders are intellectually incompetent. Or that those who successfully hoaxed them have incorrect views or are terrible people. That debate, such as it as, is an ongoing one. Indeed, one persistent feature of ongoing debates is that they quickly reach an impasse over the primary *arguenda*—the things to be argued about—and so they level up. Thus the meta-argumentative move of many of the participants. According the instigator of the meta-argumentative take on the hoax of straw men, the target view or arguers in question are so deficient that they are unable even to distinguish an actual argument from a fake one. To some extent, of course, the other victims of the hoax—the *National Review* writer—are also implicated. One might think this would be the end of the strongly meta-argumentative possibilities for the straw man fallacy: it plows right through the puzzle of effectiveness by having the victim complicit in their own straw manning. Yet there is one more strongly meta-argumentative card to play, and it's a trump card: one can do away with serious views altogether such that any attempt at critique, even pastiche, will fail because there really is no view to misrepresent in the first place.

6.4.6 Self-satire

A fundamental requirement of critical engagement is that there are stable views to be evaluated. The meta-argumentative motivation of straw manning arises from the perfectly legitimate and argumentatively necessary need to weigh contrasting reasons. And contrasting reasons are foundational to the rational preferability of a case. Weighing the contrasting reasons is of course indexed to context and can take many forms. Sometimes we're required, for example, to make up contrasting reasons when they're not presented with an objecting audience, but we just need to think of *what else* we are missing; other times we have to sweep the decks of less-than-adequate views; still other times we have to improve contrary views and look to places we've perhaps overlooked. The one thing that the contrastive program cannot do, however, is answer for views that are not there. Views that aren't really views at all. The key here is that the absence of a real, stable view, or the constant refusal of an interlocutor to have stable views might be the only thing that makes straw manning impossible. There's nothing to misrepresent. In a sense, then, this is the ultimate weaponization of the straw man fallacy since, at the very least, to be straw manned you have to have a view. But *if your view is only that you're being misrepresented*, then we might have reached the top rung of the meta-argumentative ladder.

Unsurprisingly, we see something of this phenomenon with former president Trump. The strategy is similar to the case of diving, described a few pages back. There, the point was to accuse opponents who accurately represented his words of straw manning him in order to paint himself as the victim of the injustice of being straw manned. The current version of the self-straw man takes this strategy a step further. It runs: all of my views (*wink wink*) are a joke, or aren't they? So there is nothing either (a) serious or (b) stable to criticize. This was the problem in 2016 during the Presidential campaign, as many journalists asked how they should interpret Trump: *literally* or *seriously*? (Multiple newspapers and political opinion pages asked the question with exactly those words.[17]) In any case, the problem was consistently that if the media took him *literally* and criticized him for what he had said, his defenders would say he's using some bombastic rhetoric and this literal business was, well, too literal. Trump's got big ideas—you should take him more *seriously*. And then if they took Trump *seriously*, the big ideas that seemed to be on the other side of the statements seemed objectionable, too. So they would object to the big ideas. And then the critics would ask why these critics were suspecting such earthshaking ideas out of Trump—why don't they just listen to what he says… take him *literally*. He's a businessman, and so has very concrete proposals. He doesn't work in abstractions, but in real actions and policies. And, so, on it went. Certainly, this provides the opportunity for defenders of Trump to criticize others for being unfair, but it also makes the critic appear to be themselves the victim of a kind of deliberate misrepresentation. In fact, over time, it seemed that part of the plan was to make an overall case against the press and anyone covering Trump. Only this time, unlike the "Hoaxes," the critics are provided with incoherence, ambiguity, or double-talk and then dared to make sense of it. Crucially, if they do make sense of it, they have fallen for a hoax and are thus worthy of derision, as *they cannot tell the difference between what is serious and what is not*. Alternatively, if they do not criticize it, then the instigator evades critical scrutiny.

This tactic of purposeful self-satire and buffoonery has a dark side. Jeet Heer, writing in the *New Republic*, points to a passage of Jean-Paul Sartre's *Anti-Semite and Jew* that illustrates this very point:

> Never believe that anti-Semites are completely unaware of the absurdity of their replies. They know that their remarks are frivolous, open to challenge. But they are amusing themselves, for it is their adversary who is obliged to use words responsibly, since he believes in words. The anti-Semites have the right to play.
>
> They even like to play with discourse for, by giving ridiculous reasons, they discredit the seriousness of their interlocutors. They delight in acting in bad faith, since they seek not to persuade by sound argument but to intimidate and disconcert. If you press them too closely, they will abruptly fall silent, loftily indicating by some phrase that the time for argument is past.[18]

This explains some of the apparently only half-joking strategies of certain ambiguously racist internet trolls. They use the "OK" finger gesture or the Pepe the Frog symbol and claim they're only making fun of people who think they're racist symbols. Members of the Boogaloo Movement, a far-right militia group in the United States preparing for a coming civil war based on race, which they call the "Boogaloo" invoking a movie about break-dancing (*Breakin' 2: Electric Boogaloo*), wear Hawaiian Shirts to signal their solidarity for when they face what they call "the big Luau." It's all a joke, they say. But they are carrying lots of guns, too. And they do have a habit of saying plenty of racist stuff. But they continue to play as though it's a long game of irony and antipathy toward anyone who takes them wrong when they take the bait. Helen Lewis, of the *Atlantic*, writes:

> But that's not right—bigotry and absurdity have long been intertwined. The Ku Klux Klan adopted a deliberately ridiculous name, and Klansmen claimed that they came from the moon, the historian Elaine Frantz Parsons writes in *Ku-Klux*. They endeavored "to portray victims' entirely rational fear of their physical violence as though it were superstition or gullibility. The victim, tellingly, failed to "get the joke," allowing himself or herself to be frightened by "ghosts" or "devils." The pattern repeated itself in the 20th century. "The Nazis were dedicated trolls who weaponized their insincerity to take advantage of liberal societies ill-equipped to confront them," as my colleague Adam Serwer put it.
>
> ...
>
> Extremists "rile people up by making these symbols and then denying that there's anything racist about them," Evans told me. "The goal is to make people who are actually watching out for this shit look like they're crazy to folks who haven't been paying enough attention to this."

Though popular among them, the strategy isn't only available to racist extremists. Donald Trump is known to claim that he was joking in response to criticisms of his views. Again, the famous line we'd discussed earlier about Trump seems even to have confirmed this approach: "people take him seriously but not literally."[19] His well-documented blizzard of lies and half-truths is understood by his followers, but it is a source of deep frustration for critics. Of course, a certain quotient of literalism is required if we are to evaluate views. And depriving critics of that opportunity, or what is very similar, sabotaging that opportunity by dishonestly, inconsistently, or incoherently representing yourself as doing something else cuts right to the core of argument. To some extent, this is akin to the Putin-inspired "fog of unknowability" where one tells so many lies that it is impossible to discern the truth.[20] Discerning the truth, as we've noted, requires contrast cases. The fog of unknowability exploits this, because in a fog, it's unclear what is what it is and what is what it isn't. Fog makes everything look the same. So, Trump's frequent chants to lock up political enemies are not serious—or rather, they are serious, but they are not, of course, literal. And this leads us to the final point on this ascent of meta-argumentative straw man argument. President Trump famously asked whether injecting bleach or exposing the trachea to ultraviolet light might resolve a case of Covid-19. Reporters, afterwards, asked him whether he stood by those statements. He responded that they were merely him being sarcastic:

> Q: But just to clarify—just to clarify that, sir: Are you—are you encouraging Amer-—you're not encouraging Americans to ingest—
> The President: No, of course—no. Of course.
> Q: disinfectant?
> The President: *That was—interior wise, it's said sarcastically. It was—it was put in the form of a question to a group of extraordinarily hostile people, namely the fake news media.*
> Okay. So—
> Q: Some doctors felt they needed to clarify that after your comments.
> The President: Well, of course. All they had to do was see it was—just, you know, the way it was asked. I was—I was looking at you.
> Q: No, you weren't, sir. I wasn't there yesterday. (Laughter.)
> The President: I know. I know.[21]

In a strange kind of way, this sort of meta-argumentative strategy is one that is about evacuating one's own views entirely for the sake of just inciting one's opponents. The parallel to *diving* is close, as it is an attempt to initiate contact with a defender and draw a foul for the sake of the penalty. But in this case, the program is less about promoting any view or other in the end, but in making oneself look good and one's opponents look bad. Harry Frankfurt had observed about *bullshitting* that the bullshitter doesn't care about the truth of what he says; he only cares about how *he's seen* by those listening (2005). Bullshit is a matter of, really, turning one's back on the truth or commitment to much of anything but one's self-promotion. And these meta-argumentative strategies are the bullshit of arguments about argument. But, in these cases, they aren't just making the speaker look good—they're about making the target look bad.

6.5 Taking Stock

The dialecticality and meta-argumentation puzzles are connected perplexities regarding the straw man. We are criticizing bad arguments and irresponsible views, and when we do so, we've come to reason about reasoning. How can all that be fallacious? Again, we've shown that there are instances of straw manning that are not fallacious precisely for those reasons—we may clear the decks of widespread or tempting errors, and we may come to make our own reasoning about the reasoning of others explicit. The fact that there are salutary instances of straw man arguments is, really, a good way to see why the puzzles are what they are. This is because once we see the puzzles properly, we have to refocus on what makes the straw man a fallacy to begin with. Again, we've argued that it's the inappropriate closing of an argument or critical discussion that makes the argument type fallacious. And the key is that this closure is on the basis of a meta-argument about the dialogue—that the alternatives and their representatives are not worth further attention, so one may move on.

What's particularly useful about the approach we are advocating here is that we've taken there to be an element to dialogue that's not just about what we're discussing on the first order but there's a meta-argumentative negotiation at a second order. And there is a class of fallacies that occur on that second order, when we're reasoning about the argument so far. Meta-argumentative fallacies are not just interesting but they occur because there are unique kinds of errors one can make when one reasons about reasoning. And there are many errors of this kind: from the straw man, we can see this pattern occur with the errors of bothsiderism, whataboutism, and the free speech fallacy. We are keen to explore this domain of fallacy further, but our focus now is on the most distinctive member of this class, the straw man.

The meta-argumentative nature of the straw man fallacy is particularly trenchant in instances of what we've called *strongly meta-argumentative* straw manning—cases wherein the vocabulary of critical thinking is used to construct the straw man. And, ironically enough, even the concept of the straw man fallacy can be used to construct the straw man of the opposition—that they straw man a view, and so are argumentatively vicious (and consequently, have little of value to contribute). Once we see the meta-argumentative angle to the fallacy, then we can make sense of the heated hoax literature and the cases of what we've called *argumentative diving*. We are reflective creatures, so it's not a surprise that we reason about our reasoning and we argue about our arguments. And since we are also fallible creatures, we err when we reason. And we have tools to correct those errors. But this does not eliminate our fallibility, and so we err with those tools to correct our errors. And so, we err when we argue about arguments. That's how we get meta-argumentative fallacies, and why they need a special class of attention and evaluation.

7

Consequences for Fallacy Theory

We have argued for a number of things up to this point. Here are some highlights: the straw man triad of the representational straw man, the selectional weak man, and the hollow man. There is a related version of the straw man fallacy, one posited making an interlocutor's argument stronger, instead of weaker, the iron man. Straw man arguments can be argumentatively salutary (and that iron man arguments can be fallacious). And we've tried to answer the puzzles of effectiveness, dialecticality, and meta-argumentation. And out of the answers to these puzzles, we've developed some theoretical and explanatory views on the straw man. That there are different forms of address for arguments, and that the effectiveness question should be indexed to those forms of address. That there can be meta-argumentative fallacies. And that some concepts of reasoning have strange reflexive fallacious instances—what we'd called *strongly meta-argumentative fallacies*. We believe it is clear that the straw man is a fecund topic for argumentation theory and that there is much to take from what we've shown. Our concluding act here comes in two parts. First, we will turn to some of the standing literature in fallacy theory on the straw man to answer some challenges to our view and to show how we can correct some problems other theoretical programs face when accounting for the straw man. So we are out to update the state of dialectical play in the fallacy-theoretic literature on this fallacy. Second, we think that a well-theorized fallacy provides information for progress in fallacy theory and broader argumentation theory. We think we have identified a few fronts for development. They are, on the one hand, that audience-indexing is an important element to understanding argument-development, and audiences are not always dialogue partners. The onlooking audiences to arguments are important variables for our interpretive and evaluative programs. So the dyadic model of dialogue is incorrect—a polyadic model is necessary to properly theorize the variances in critical argument. On the other hand, once we see the straw man is not only a dialectical fallacy but a meta-argumentative fallacy, it is worth theorizing what related meta-argumentative fallacies there are. And if we are right that fallacy theory itself makes a particular class of what we've called *strongly meta-argumentative fallacies* possible, the question is: What is our best way forward?

7.1 Current Fallacy Theory and the Straw Man

As we proceeded with our own case in the previous chapters, we regularly referenced a wide variety of other theorists with whom we've been in critical dialogue on the straw man fallacy, and most of these references have been in passing. And we completed our history of the development of the view of the straw man as a unique fallacy with a look at its uptake in contemporary textbooks, but we left the current scholarly discussants of the fallacy out of the completion of that history. Our primary reason for these presentational choices is that we've tried to highlight what we've owed to these scholars in the presentation of our own view, and we believe that with the theoretical edifice we've assembled, we can more effectively discuss where we agree and break with our contemporaries on the theory of the fallacy. For sure, we editorialized liberally in presenting the history as it was, but given the current controversies over the fallacy and the explanatory apparatus for theorizing it, we thought it best to have the exchange with our contemporaries presented more in the form of objections and replies here at the end of the book. Again, our objective is to situate the view among the main theoretical approaches to the fallacy and within fallacy theory generally. Our overall case will be, perhaps unsurprisingly, an argument from contrasts. The contrastive case will be that our cases represent a useful variety of forms for distortions possible for straw manning that require their own explanatory and corrective tools, that our approach handles the effectiveness, dialecticality, and meta-argumentative puzzles for the straw man, and that it represents a fecund way forward for further work in fallacy theory and argumentation theory generally. And it does better on all of these fronts than its competitors. Fecundity, for sure, is a pragmatic criterion, one that depends on what one thinks are important or interesting topics for future work. So it has a bit of an *eye of the beholder* quality to it, but we will make the case later for the meta-argumentative consequences of our approach being of real significance beyond the straw man. Meta-argumentative fallacies themselves require that the arguments under consideration are properly theorized as not only about what is up for discussion but how we've discussed the discussion, and given that much of our public discussions involve discussions of how we discuss things, it's important that we have clarity on how and where we err when we reason about our reasoning. That, we think, is not just interesting but important. The other two criteria, capturing variety (as in, approaching completeness with the phenomenon) and explanatory depth with the puzzles, are more objective criteria. The key with explanatory depth is that we need, again, to keep two things in mind: we need to explain both what's fallacious about the straw man fallacy and how it works. One point that will emerge is that most theoretical programs that are good at one horn of this requirement fail at the other. That is, the better a view is at explaining why the straw man is fallacious, it often is more difficult to see how it could be effective. Or the better it is at explaining how it's effective, there is a question of how it's fallacious. Our program, again, is set up to explain both. We will survey the main competitors to our view in what follows, and we will show that these contrasts all cut in our favor.

By our lights, there are roughly four competing approaches to the straw man fallacy in the going literature. First, there is the pragmatic program, represented by Douglas Walton's work (1996, 2013). Second, there is the pragma-dialectical approach, represented by Frans van Eemeren and Rob Grootendorst's account in *A Systematic Theory of Argumentation* (2004) and work following in the pragma-dialectical tradition, as seen by Lewiński (2011, 2020) and Lewiński and Oswald (2013, 2014). Third is work in the pragmatic inferential tradition, by Louis de Saussure (2018). And fourth, and finally, there is the rhetorical-audience indexing program, represented most prominently by Christopher Tindale's account of the role of audiences and the problem of the straw man (1999, 2004, 2007, 2015). We will survey these four programs and make the case that our account outperforms them along the variety of explanatory criteria. To close this section, we will discuss how our approach dovetails with ongoing empirical research on presentational variables for effectiveness for straw man arguments, as seen in the work of Bizer, Kozak, and Holterman (2009), and Schumann, Zuffrey, and Oswald (2020, forthcoming).

7.1.1 Walton's Pragmatic Account

Douglas Walton identifies three elements comprising the straw man fallacy. First, what he calls the three-part structure of reasoning that he approvingly notes was first accounted in Johnson and Blair (1983: 70):

Two arguers, M and N, with two positions to consider, Q and R.
1. M attributes Q to N.

2. N's view is not Q but R.
3. M criticizes Q as though N holds it.

Second, that the straw man "interferes with the basic goal of a critical discussion, and is therefore at cross purposes with this type of dialogue." And third, that the straw man needs an account of it being "a distinctive species of fallacy in its own right," namely that "the essence of the deception or error inherent in the straw man fallacy [makes it] a distinctive type of sophistical tactic" (1996: 125). Walton's core analysis of the fallaciousness of the straw man is that it resides in the *misrepresentation.*

> The precise reason why the straw man is normatively counterproductive in critical discussion is that for the critical discussion to proceed in resolving the conflict of opinions by reasonable argumentation, it is necessary that each party argues against the other party's side by using premises that the represent the commitments (position) of that other party. Otherwise, the dialogue is at "cross purposes."
>
> (1996: 125)

Walton emphasizes that "failure to engage with the real position of your opponent… defeats the whole purpose of your argument" (1996: 121). Walton later takes the same route, noting that the straw man fallacy is a dialogical problem entirely in the fact that there is no fit between the move of one discussant and the one previously by the other, and this failure of fit is in the "distorting of his previous messages" (2013: 250).

Importantly, Walton is keen to emphasize that the straw man is a uniquely dialogical fallacy—it needs one arguer attacking another. And so he notes:

> The interactivity of such attacks that are immediately denounced shows that the straw man fallacy requires a dialogue model for its analysis. One cannot merely analyze the argument as an isolated set of premises and conclusion. Instead, *the sequence of moves in a dialogue between two parties needs to be taken into account.*
>
> (2013: 250 emphasis added)

So, as we see it, Walton has taken on a dialogical perspective on the fallacy, which we endorse, but his approach is explicitly *dyadic* in form for the dialogue, referencing only *two* relevant dialogue participants. This dyadic program immediately faces a challenge, what we'd called the *puzzle of effectiveness,* when Walton turns to explain how the fallacy works in practice and how it can then be challenged. He notes that "if the respondent is actually present when the charge of fallacy is to be evaluated, then the case is quite different from the situation where he is not able to comment" (1996: 126). If present, "it may not be too difficult to reply… by insisting that his position is not what the proponent has pictured it as," (126) and "if the respondent is not present, then the evaluators should be required to go very strictly by the existing discourse…" (127). But this is a strange way to be committed to the model of there being *two* in the dialogue. In short, it seems to *concede* that if the straw man is presented in the second person (to the target as *you say*…), it will be caught and be utterly ineffective. And if it is presented in the third person (about the target to an audience as *they say*…), then it requires a *third* participant in the dialogue, the "evaluators" to detect the misrepresentation or not.

Given the way we've formulated the effectiveness puzzle, *how can the straw man work as an argumentative tactic when posed to their targets?*, we think Walton has no answer. He either says they don't work or he says they aren't posed to their targets. The first route makes us ask how it could be a fallacy worth theorizing (why theorize a fallacy that's impossible to get away with, and why have it on a list of fallacies to keep an eye out for?), and the second route shows that Walton's dyadic model for critical dialogue is incorrect—we need a *triad* of speaker, target, and audience to make sense of what happens there. Either way, Walton's program underperforms on the core puzzle of the fallacy—how in the world could it be effective? Again, our answer is that in the second-person form, the fallacy could either be a silencing epistemetric tactic of making the exchange too costly for the target to continue or as a form of gaslighting the target into thinking they have weaker views than they originally thought they'd had. And in the third-person form, the answer was that we needed a *polyadic* model for critical conversation—we need a variety of arguers, audiences, and so on, but most importantly in this case, an audience who doesn't detect the distortion or who actively cheers it out of the hermeneutics of antipathy.

An additional problem for Walton's program is that he locates the fallaciousness of the straw man in the *misrepresentational* element of the tactic. "[T]he straw man fallacy is rightly seen as a fallacy of misattribution of commitment" (2013: 251). But, as we've shown, this is too quick, as there are misrepresentations that can be argumentatively salutary. There are virtuous straw man misrepresentations of opponent's views that help clear the decks of common errors, there are straw men that impel further clarifications, and there are pedagogical straw men that allow for the development of one's argumentative skills. They are not fallacious by our lights, even though they misrepresent their targets or the current state of the critical dialogue, but they are salutary because they have long-term or *diachronic* positive effect on the critical discussion. The fallacy of straw manning is not in the misrepresentation but in what we've called the meta-argumentative closing function on the dialogue. That with this bad representation of the target, we infer that there is no more to say to or expect from them on the matter. And we prematurely close the argument.

Walton has criticized our approach. His primary criticisms target our triad of representational standard straw man, the selectional weak man, and the hollow man. As we take it, his critical line here is to deny that these forms are so distinct that they deserve separate treatment, and so our claims to theoretical fecundity with the variety of kinds (and now to *burning man* and *iron man*, too) is unfounded. Walton's argument is that both the weak and hollow man are really just standard representational straw man arguments and that charges of weak manning have a problem with provability. We will start with Walton's case against the weak man as having a provability problem, and then we will turn to the hollow man issue.

Walton's primary challenge to the weak man is that if there really are bad versions of a view or argument out there, then they not only should be fair game for a critic to select them but the charge of straw man for the one making the charge of weak man here would be harder to make than with those in representational form.

> What Aiken [*sic*] and Casey call the weak man fallacy is problematic, precisely because of the dialectical variability of argument.
>
> (2013: 281)

> In short, just because the proponent looks over several arguments put forward by the respondent to defend his or her position, and selects the weakest of these as the focus of his or her critical questions and counterarguments, it by no means follows that the proponent has committed the straw man fallacy. It may be a good strategy.
>
> (2013: 283)

There are a number of things to object to here, from our perspective. The first is that we don't hold that it is in the selecting the weakest arguments for refutation that the fallacy of weak manning resides. We hold that it is in the closing meta-argumentative function of weak manning that the fallacy resides—that the weak arguments selected are representative of the best the other side can muster, and so the debate is practically finished. In addition, it is *our point* that weak manning can have argumentatively salutary consequences. If Walton's view is that the fallacy of the straw man is in *distortion* of the other side, then this kind of nutpicking opponents on *his theory* should be vicious. It's worth noting, again, that the weak man strategy, if deployed as representative of the best of the opposition, is not, as Walton holds, "a good strategy," since on Walton's theory, such a misrepresentation "defeats the whole purpose of your argument."

The second issue is with the *provability* of a weak man charge. Walton's challenge to us is that given the "dialectical variability of argument," such charges of weak manning do not stick, because showing that the other side chooses a weak version of a view when stronger ones are available, there is a matter that would essentially chose sides or not allow critics to go after an argument they see in need of criticism. That is, because there's a disagreement about an issue, there's a disagreement about what the best case on either side is for resolution. And further, we shouldn't prohibit arguers from going after whatever arguments they see as worthy of criticism, especially the bad ones.

> It needs to be up to the proponent to make a decision on which of these arguments to select, because it is an important feature of the critical discussion, or persuasion dialogue, that each party be allowed the freedom to probe into the position and supporting arguments of the other side in order to find them weaknesses in them that most need to be critically scrutinized.
>
> (2013: 283)

So a speaker should be *free* to target whatever arguments they like for criticism. But this point from Walton is irrelevant to the point about the weak man as a move in contributing to critical dialogue overall, and as a fallacy in that exercise. We are not saying that critics shouldn't be free to criticize wherever they like in the opposing side's case. In fact, we've argued explicitly that going after bad arguments on the other side can be argumentatively salutary. Further, the question is not about that freedom to pursue whatever argument that arguers see worthy of critique but rather about what they've shown and what they purport to have shown with selecting only the weakest arguments of the opposition and leaving the strongest alone. That's the fallacy of the weak man, again, not that they criticize weak, instead of strong, positions of the opponents. In essence, Walton's defense here is to invoke a freedom norm to license the move and criticize our case. But we've not prohibited it. We've argued that certain free critical lines of argument that only go after the weakest arguments don't show much, or that pretending that one has refuted the entire case of the other side by only going after their weak arguments is the problem. In fact, we've argued that there can be argumentatively salutary instances of this weak manning move, under the right conditions and with clarity about the restricted results. Walton's point about freedom norm here is just irrelevant. What matters is *what follows* and *what the arguer takes to follow* from weak man arguments. Our point was the meta-argumentative inferences that the weak man licenses or not is where the fallacy lies, not in whether arguers are free to argue as they see fit. Surely there's a name for the error of argumentative attribution here.

A final note about Walton on the weak man is that dialectical variability speaks to the fact that the critical dialogue cannot be dyadic in the relevantly restricted sense. There are many contributors to debates, and sometimes the weak man is not about selecting one argument from the other arguments a singular opponent in the dialogue gave but in selecting the weakest arguments and arguers from the many discussants in the dialogue. The weak man, we think, highlights why the dialogue model must be polyadic, not dyadic.

Walton's criticism of our concept of the *hollow man* is short and crisp. He thinks that the hollow man is just a special instance of straw manning.

> The hollow man fallacy is not a problem so long as we define it more narrowly as a version of the simple straw man fallacy that is a more extreme variant in which the position attacked by the proponent is a mere caricature that is so far removed from his or her real position that it bears no relation to it at all.
>
> (2013: 285)

Walton's case is that the representational straw man and the hollow man are different only in *degree*, not in *kind*. "Seen in this way, the hollow man is a subspecies of the simple straw man fallacy" (285). Our reply is that *misinterpreting* one's opponent and *fabricating* what they say *is a difference of kind*. One way to see this is in terms of the burden of proof that straw man accusations of the two kinds have. One points, say, to the record for statements from the target of the straw man and shows that statements X, Y, and Z were taken out of context, manipulated, or left unqualified. That's representational straw man and how one unmasks it. But there is nothing to point to with hollow manning, because it is not a re-presentation of what the target has said, as it doesn't reframe or distort the target's on-record statements, because there is no reference to them at all. So, consider "Bush and Iraq" from Chapter 3. When Bush attributes racist views about Iraqis to his Democratic critics, there were no people named, no statements referenced, and no general forum where statements like those may have gone on record. So of whom and of what is Bush speaking? But Bush attributed them to his opponents, and the response isn't then to go look over the newspapers or talking heads shows to find what he misinterpreted. Rather the response was to say, *Nobody said that or anything like that*. The tools of reply and evaluation are different between the two here, and this, we think, makes it clear that the hollow man fallacy is different.

Finally, it should be clear that, as stated, Walton's account of the hollow man as an instance of the straw man in dyadic form is a *hopeless* argumentative strategy, as it suffers from a significant problem of the puzzle of effectiveness. Returning to "Bush and Iraq," surely Bush didn't think he was correcting *the Democrats* as he made his hollow man of them, but that he was doing it for his Republican compatriots. So, again, our answer has been that hollow manning, and particularly the Bush case, shows that much critical dialogue is at least triadic between the speaker, the straw man target, and an audience who either cannot detect the distortion or are active participants in the hermeneutics of antipathy directed at the target. The hollow man is just not produced for the target.

7.1.2 The Pragma-dialectical School

According to the pragma-dialectical school of argumentation theory, the straw man fallacy is fallacious because it breaks a rule of the code of conduct for reasonable discussants. In particular, it breaks a rule that van Eemeren and Grootendorst call *Commandment 3, the Standpoint Rule:*

> Attacks on standpoints may not bear on a standpoint that has not actually been put forward by the other party.
>
> (2004: 191)

The rule, as stated, is to ensure that criticism and defenses "really relate to the standpoint that is indeed advanced by the protagonist." The pragma-dialectical rationale for the rule is that real resolution cannot be achieved if the speaker criticizes a view the target does not hold. The rule "ensure(s) that the attacks and defenses carried out in those parts of an argumentative discourse or text that represent the argumentative stage of a critical discussion are correctly related to that standpoint that the protagonist has advanced" (2004: 192). And, as noted by van Eemeren, Grootendorst, and Snoeck Henkemans:

> If parties talk at cross-purposes like this, it will be impossible for them to resolve the original disagreement. Even if the dispute seems to be resolved, it will be, at most, a spurious resolution.
>
> (2002: 117)

Given that critical dialogues are, on the pragma-dialectical model, instances of strategic maneuvering of the sides to prove their case within the rules, Marcin Lewiński identifies the problem as "derailed strategic maneuvering," as the arguer has "misrepresented his opponents position for an argumentative gain" (Lewiński 2011: 476, 2014: 207–9; Lewiński and Oswald 2013: 165; Oswald and Lewiński 2014: 329 and noted in passing much earlier by Krabbe 2002: 158). The pragma-dialectical program provides an ideal dialectical procedure for resolving disagreements on the merits of the sides by strategic testing from representatives—and the straw man is an instance of the standpoint rule being broken. Further, and argued by Lewiński, the pragma-dialectical relation is *dyadic*; there are two disagreeing participants. So even detecting and criticizing a straw man requires the target on the scene.

> The actual presence of original arguers is dialectically vital, because only then can they in the course of discussion correct a possible distortion.
>
> (2011: 481)

But this "disagreement space" has some tolerance for degrees of representational mismatch between target's statements and what is criticized, since there must be room for strategic maneuvering with interpretation on the side of critics. The principle of charity has exceptions in these spaces, since arguers should be allowed "plausible interpretive maneuvering" as a strategy of easiest objection (2011: 493). So long as the interpretation of the opponent is not both "implausible and uncharitable" for the context of the exchange, some representations *can* be acceptable, even if not perfectly accurate or charitable (Lewiński and Oswald 2013: 172).

We will start by noting a number of places where we agree with the pragma-dialectical program, as we owe much to it in developing our own. First, identifying the problem of straw manning as breaking a procedural rule of discourse seems an insightful strategy, and the insight that it is a result of adversarial relations between speakers in that rule-bound domain is, we think, exactly right. In addition to this, Lewiński (2011, 2020) and Lewiński and Oswald (2013) have made great progress in identifying a tolerance zone of appropriately adversarial (instead of maximally cooperative and charitable) interpretive comportments between speakers that reflect the gritty reality of disagreement. So we agree there are instances of breaking the standpoint rule as stated. Some misrepresentations of dialectical opponents are not only allowed but argumentatively salutary. In fact, given this point, it perhaps could be said that the standpoint rule should be restated for the pragma-dialecticians:

Attacks on standpoints may not bear on a standpoint that has not actually been put forward unless that standpoint is neither implausible nor uncharitable of the other party's contribution.

Regardless of whether Lewiński's cases are revisions of, or exceptions to, the standpoint rule, there is room for what we've called salutary straw men, and the case for them is primarily in their diachronic function of contributing to the dialogue as contest.

A final point worth noting before we turn to our critical account of the pragma-dialectical school on this fallacy is that Oswald and Lewiński are the only theorists who spend a good time on the fact that the straw man has a significant *meta-argumentative* element. By their lights, the straw man is an operation on the discourse of an opponent by strategic reinterpretation and criticism (2014: 313), and the bounds and motives for fallacious instances (i.e., those both uncharitable and unreasonable) are those fed by a "bias-generated error" (2014: 315). Straw man arguments, then, paint a picture of one's opponents and make that picture the cognitive priority.

So there is much we think we have in common with the pragma-dialectical program as it stands on the straw man fallacy. We agree on dialecticality, the meta-argumentative nature, and the fact of non-fallacious cases. But there are many points where we disagree. The first and primary point is that, given the dyadic model for critical dialogue for the pragma-dialectical program, there is always the puzzle of effectiveness for fallacious instances. In fact, when van Eemeren, Grootendorst, and Snoeck Henkemans note that straw man fallacies "make it impossible" to resolve the issue, they simply restate the problem for themselves. How can straw man arguments work at all on this model? Lewiński's work is devoted to answering the matter in a way, which is, again, to identify a range of non-fallacious misrepresentations. So that explains how *non-fallacious* straw man works. That's progress, *but that still leaves the fallacious instances to be explained*. How in the world do *they* work on the target of the straw man? And our diagnosis of the problem for the pragma-dialecticians, again, is the inherent dyadic model for the pragma-dialecticians. In fact, it comes out even when Lewiński articulates how to make the case that an arguer has straw manned. He says that only the arguers themselves can correct a distortion—"the actual presence of the original arguers is dialectically vital" (2011: 481). That is a form of *hyper-dyadism* about critical dialogue, since it follows that *nobody else can identify when you've been straw manned*. That seems wrong, since anyone who has taught a class in the history of philosophy knows that calling out straw men of the long dead posed by our sophomores is the main task of the job—how many inappropriate misrepresentations of Epicurus can a room of college kids come up with? Surely we don't have to get out the old Ouija board to *really* correct them, right? Lewiński is right that it *helps* to have the speaker there, but he's wrong that it's necessary. Now, to his credit, Lewiński (2014) has challenged the dyadic model of the pragma-dialectical dialogue. He imagines a kind of poly-logical straw man where the fallacy consists in confusing the standpoints of real opponents. In this case, there are three conversants, A, B, and C. And A straw mans C by attributing B's objectionable view to C and refuting it. Lewiński notes that this is "a faithful attack on a central argument of a real opponent—just not precisely the right one" (2014: 208). He calls this a "*collateral* straw man." We think this is a salutary development in the theory, and we think this kind of confusion in collective argumentative contexts needs a good name, and this is excellent. But, crucially, the poly-logical scenario is a dialogue involving (in this case) three speaking *participants*, all of whom are *present and speaking* (208). Again, the overall issue is that the pragma-dialectical program is one devoted to an argument match, but argument rarely comes that way. There are many conversants, onlooking audiences, and opponents who may back out of the argument for reasons that are purely practical—time, energy, patience. Or they may be long dead. And, again, the effectiveness puzzle isn't answered with this line of polylogue, since the straw man of C won't convince C of anything. But on our model, we have an explanation—it convinces the onlooking audience to the exchange.

Our second disagreement with the pragma-dialecticians is on the breadth of non-fallacious straw men and what makes the form fallacious. We think there are non-fallacious versions of the straw man that are, on the Lewiński-Oswald metric, implausible and uncharitable interpretations of the view. "Hamlet's Ghost," from Chapter 4, is a case of posing an implausible and uncharitable interpretation of having the ghost on stage (as "stomping around with a bed sheet over his head"). But, if this challenge is performed under the right conditions, it can prompt further thought on stagings, how a ghost can reasonably be dressed and play a role, and so on. It can be posed as a *spur* for critical dialogue. The same goes for "Religion and Morality" in Chapter 3, as the whole point of performing the argument was to be uncharitable to implausible views so that they are excused from the discourse space in order for a well-run argument to proceed. The takeaway is that the pragma-dialectical program still takes the fallaciousness of the straw man to be in the misrepresentations, and the Lewiński-Oswald revision was to give this criterion a penumbral edge. That, we think, is progress, but it is still erroneous. The fallacy of the straw man, instead, lies in the meta-argumentative move of prematurely closing dialogue with one's opposition. There are instances where the misrepresentation contributes to maintaining and even improving open dialogue. That's what makes salutary straw men.

7.1.3 de Saussure's Pragmatic Approach

Louis de Saussure's approach, we think, is the only that properly appreciates what we've called the effectiveness puzzle. The straw man approach, on de Saussure's line, is an attack on the target's argumentative skills. So de Saussure distinguishes between two questions regarding the effectiveness of fallacies generally (and the straw man, in particular):

1. How is it that fallacies are successful in persuading?
2. How is it that fallacies may occasionally fail in persuading an audience but still constitute a winning move in some argumentative interactions? (2018: 172)

One of the key thoughts here is that winning an argument is not identical to eliciting assent or gaining adherence. One can win an argument without yielding conviction. And de Saussure also notes that the fallacy "generally involves a third party" to judge there being a success. He notes that there is a meta-argumentative dialectical core to the fallacy—that it, ironically, is a fallacious accusation of fallacy. The core of the straw man, then, is that of being merely a "pragmatic winner" with the target. The speaker straw mans the target, and if it is in the second person, the burden for challenging the misrepresentation is complicated and meta-discursive. It requires a good deal of skill for the target to detect, reflect, and correct this kind of error. So, if the target is not skilled enough to do so, the author of the straw man "gains prestige by exercising and showing his pragmatic, cognitive skills" (2018: 185). And the targets are downgraded in these estimations, even if the targets themselves do not find that the critique bears on the views actually held. Thus, the straw man yields "pragmatic winners" (188).

We believe this program is very insightful, particularly in addressing the effectiveness puzzle in the hardest instance, the second person. And for distinguishing between winning a dialectical exchange and yielding conviction from the exchange. So we agree with the program, framed as it is, and it should be clear that we take de Saussure's line of explanation to bear on what we've called the "purely dialectical" silencing version of second-person straw manning. Though it seems it could, if the status issue is one that is given *for the target*, also be part of the groundwork for the gaslighting version of second-person straw manning. On the purely pragmatic winners' account of second-person straw man, the straw man is a form of gamesmanship in the game of giving and asking for reasons. It is like a player who hacks other players over the course of the game—the fouls make the game harder to play at a high level, and they accumulate to injury that may prevent play altogether. So one can win the game, but not convince the other side that they have been truly outplayed. But for as deep as this line of argument de Saussure provides, we think there are significant points of divergence between his and our view.

The first is simply that the model explains the second-person straw man, but it requires that the target for the straw man be present for the presentation. This explanation just can't work if the target is not present. We don't get the downgrading of their skills if they aren't there to flail and fail with addressing the straw man. That is, on de Saussure's model the target must be present to be dunked on. But the straw man regularly happens when the target is not present. They may just not be there at the time of the representation but elsewhere, or they, again, may be dead. Do we really think that a sophomore who straw mans Aristotle on existential assumption for their midterm paper in Ancient Philosophy is in a good position because Aristotle doesn't answer? So de Saussure's explanatory program is good for what it's for, but it's not good as a general explanation. It is still a dyadic program, and it is, like the pragma-dialectical theory, a kind of hyper-dyadism, since for the straw man to work, the target has to fail with the reply for the relational pragmatic results to follow.

Further, it is not entirely clear from de Saussure what makes the straw man fallacy fallacious to begin with. Part of de Saussure's explanation for how it is a winning dialectical strategy is that it "reverses the burden of proof" (2018: 176) on the straw man target to untangle the misrepresentation. But that cannot be what makes straw manning fallacious, since all criticism of others' arguments changes the valence of the burden of proof on the meta-argumentative level. That is, all critical replies to an argument are about weighing out how accurate a case or a critique is, how challenging or relevant an objection is, and so on. As we've argued earlier, all argument exchange has a measure of meta-argumentative overhead, so it can't be that straw man is a fallacy because of the meta-argumentation that must be performed by the target. And in light of there not being a clear demarcation of what makes the straw man fallacious, there is the added issue of whether there can be appropriately posed straw man arguments. de Saussure's account is not complete enough to answer these challenges. To be fair, it was only a single article devoted to a restricted version of the effectiveness puzzle, but it is not a complete theory.

7.1.4 The Rhetorical Approach to the Straw Man

The rhetorical theory of argument is the approach to argument that identifies argument quality as a function of effectiveness with reasonable audiences. As Perelman and Olbrechts-Tyteca hold, "the goal of all argumentation... is to create or increase the adherence of minds to the theses presented for their assent" (1969: 45). James Crosswhite argues that "the aim of argumentation is to gain the adherence of other people; all argumentation develops in terms of an audience .... [A] rhetoric of reason calls for a reception theory of rationality" (1996: 36). And Christopher Tindale holds that "rhetorical argumentation... has its primary concern the attempt by an arguer to gain or increase adherence of an audience for a thesis" (1999: 69). The rhetorical program on straw man arguments has been the most effective at answering the puzzle of effectiveness, because the rhetorical program, especially that of Christopher Tindale, is particularly keen on interpreting arguments in light of their audiences. So a model for our polyadic approach to dialogue has been in the works in this tradition for a long time. Arguments have a variety of audiences, depending on how and when they are delivered, and who else might be listening in or later told about the exchange. There may be an abstract and ideal *universal audience*, on the Perelmanian model, which is an arbiter of deep rationality and truth (Perelman and Olbrechts-Tyteca 1969: 31 and Perelman 1982: 24), and arguments can have *implied* audiences, even if they happen to be *in situ* addressed to others (as noted by Crosswhite 1996: 139). Tindale is careful to note that arguments themselves should be interpreted in light of the kind of audience they are intended to reach (1999: 76, 2004: 12, 2015: 212). And this yields significant dividends with regard to explaining both how straw man arguments work and how they are fallacious. Tindale argues that whatever semantic relevance that may be present or not in straw man arguments is not the salient notion of (ir)relevance that explains the argument as a fallacy. This is because the issue is about a misfire between speakers. He initially terms the problem a matter of being "contextually irrelevant" (1999: 37) and, later, "dialectically irrelevant" (2007: 24). Tindale explains that with straw man arguments, a critic does not represent the target accurately. So there is a process norm of address broken. Consequently, "A 'straw man' argument would seem always incorrect and have no redeemable instances" (2007: 12). On the rhetorical view, then, the fallaciousness of straw man arguments resides in failing to address the arguer with opposing arguments, but the effectiveness of the strategy is in appealing to *another* audience, constructed or selected by the speaker. Tindale introduces Bakhtin's idea of the *superaddressee*, a third party to the dispute, upon whom the author projects an image of properly hearing and interpreting the case. And they communicate the preferred attitudes regarding the case at hand.

> Insofar as the superaddressee represents responsible understanding, and understanding cannot be from the outside, the superaddressee is internal to the utterance. Furthermore, this superaddressee is 'presupposed' by the author of the utterance, it is controlled by the author.
>
> (2004: 126)

The point is that the speaker, in constructing a straw man of the target, must be doing so for a third party that the speaker projects as sympathetic with the portrayal. Many onlooking audiences, then, sort themselves out into receptive groups—those who agree happily continue to listen, and those who do not either opt out or argue back. On the rhetorical theory of argument, then, straw man arguments are effective because they construct audiences as they are given.

The rhetorical theory, as we see it, has a deep explanation for the effectiveness of straw man arguments and even the way they can *direct* their audiences to see themselves as standing up for what's right and be comforted in the clear refutations of their targets. And they can, in the second person, explain how the construction of an audience in the target can slowly chip away at a subject's epistemic self-reliance and sense of standing. The idea of constructing an audience with the argument is a powerful idea, for sure, and it should be clear that we've been influenced by Tindale's rhetorical program. But, we think, there are two problems for the rhetorical program. The first is one we've already noted in Chapters 3 and 4 with how we've broken with Tindale—the straw man *does* admit of virtuous and argumentatively salutary instances. Given that the account of where the fallacy of straw man resides for the rhetorical theory, in the misrepresentation, there can be no virtuous versions.

The second point of disagreement between ourselves and the rhetorical program is one rooted in the nature of the rhetorical program itself, but it can be revealed with the case of the straw man. On the rhetorical theory's identification of argument quality with audience-effectiveness, there is the benefit of interpreting arguments and assessing their quality indexed to their audiences. But this also yields unhappy results. For the audiences of effective straw man arguments, there is a rhetorical sense that it is also a *good argument*. For sure, for the *targets* of straw men arguments in the second person form they are misrepresentations and are even *unfair*, as Tindale notes (2007: 22). But if audiences assess these as fair and relevant (perhaps because they do not have more evidence of the debate or representations of the target's case beyond what the speaker provides), there must be a sense that this argument is a good one. So the virtue of the rhetorical program, its focus on effectiveness, is also its vice, since the rhetorical approach is one that identifies argument quality with the capacity to increase adherence. But the normative component must be a separate matter, because we are out to explain how bad arguments can be effective. This, Aikin has argued elsewhere (2011b), is an unacceptable result for fallacy theory and for argumentation theory generally. A reception theory of rationality will have a good deal of explanatory power for how arguments move us, but the costs will be demarcational force with normative evaluation of those arguments.

7.1.5 On Empirical Programs on the Straw Man

Here, we will briefly address two lines of empirical research bearing on the straw man fallacy, particularly on variables influencing effectiveness. We believe that these cohere well with our program.

George Bizer, Shirel Kozak, and Leigh Ann Holterman have shown that persuasiveness of straw man arguments vary with audience motivation to scrutinize the cases. In instances with high personal relevance, the subjects were motivated to higher degrees of scrutiny, and so were not convinced by straw man arguments. In cases with low personal relevance, the subjects were not motivated to be scrupulous, and so were more likely to be moved by straw men. In the Bizer cases, the tests were speeches from fictitious candidates for public office that gave straw man arguments against their opponents regarding matters of traffic safety or higher standards for food inspection. In both cases, the straw man of the other candidate was that "Some say that by passing a bunch of laws, we'll fix everything...." If the candidate is running for office in a faraway state, the straw man arguments were more persuasive than if they were offices in the state wherein the subjects lived. The reason, again, for this variance was that the subjects would (hypothetically) have to live with the consequences in the latter cases, but not in the former. Bizer, Kozak, and Holterman observe, then, that:

> [T]he technique seems to be relatively effective when people are disinclined to carefully process a message.
>
> (2009: 224)

> [I]t was ineffective... among people who had such motivation.
>
> (2009: 225)

The lesson generally is that personal motivation to be attentive and critical is important to detecting bad reasoning and not being moved by it. And without this motivation of being a stakeholder of the outcome consequences for the argument, one is vulnerable to low-quality reasons, and low-quality representation of a dialogue partner's reasons in particular.

We think that the Bizer, Kozak, and Holterman findings cohere well with our program, as stated. First, all of their cases were in the *third person*, and the subjects were able to see the texts of the arguments exchanged, so they were able to detect the distortions in the cases—they read the statements from both sides, so were not dependent on the straw manning speaker for the picture of the target of the straw man. And the point about personal stakes reducing likelihood of effectiveness is a good way to explain why it seems clear that straw man arguments have a problem when posed in the *second person*, since the *you* addressed will feel considerably more invested in the results. Further, it seems clear that the increase of persuasiveness for audiences without personal connection to the issue is a useful explanatory result, since we think there is likely room for thinking that personal stakes can make for instances of *motivated reasoning* behind a form of *self-serving bias* in argument, too (see, e.g., Baumeister 2005: 214). That is, if one's personal stakes are for the argument to come out one way or another (instead of for whatever the best view is, in the end), it seems that our preferences for how the argument should go would influence how one assesses the persuasiveness of the straw man arguments, too. Given this, we have a way of seeing how, then, escalated adversariality can make it so that it's easier to see even weak arguments against the other side score deadly points.

Jennifer Schumann, Sandrine Zuffrey, and Steve Oswald have provided compelling evidence that presentational details matter for straw man effectiveness. They identify a forewarning effect that subjective connectives play in stating a side to be straw manned. When stating another's view as the result of them drawing subjective inferences, as opposed to reporting facts, one's audience is more likely to find the former cases less convincing. So, for example, the English word "since" is ambiguous between the *subjective inferential* sense or an *objective causal* sense. Take the two following claims:

> Since he missed the train, Harry was late to work. (Objective-Causal)
> Since he was late to work, Harry must have missed the train. (Subjective-Inferential)

The point is that the second "since" forewarns listeners to a *supposition*, and so hearers are more scrupulous about evaluating what comes next (2020, forthcoming). So we have tools to explain how the representational details of straw man arguments can influence their reception beyond those of bald misrepresentation, complete fabrication, or taking statements out of context. There are more subtle ways to manipulate presentation that will make it more likely that an audience will assess a target's views less favorably.

We believe that this account shows that this fallacy is most theoretically tractable in cases of *third-person presentation*. That is, all of the cases tested were ones with what were considered the two essential features of the fallacy (a distorted attribution and an attack on the distorted view), but they were reports *of other people's views and plans*, so were in the third person. The second consequence here is that in cases of detected straw men, tested audiences preferred non-fallacious exchanges. This coheres with our point about the puzzle of effectiveness, that if the misrepresentation is detected, it won't yield acceptance (again, with exceptions of those proceeding with antipathy for the target or in second-person silencing cases). This was why we needed to invoke a third party as onlooking audience who did not know any more about the situation than the speaker tells them. As Schumann, Zufferey, and Oswald note, participants who heard more of the dialogues spotted the fallacies and preferred the non-fallacious versions (forthcoming). And again, this makes sense—straw man arguments work on audiences that depend on their speakers for pictures of the target. If a neutral audience has more information about the exchange, the distortion will be more likely detected, and the straw man argument won't stand.

7.2 Prospects for a Theory of Meta-argumentative Fallacies

Our brief history of the straw man in Chapter 2 found the straw man to arise from a misinterpretation of Aristotle's *ignoratio elenchi*. While Aristotle meant this in a very general sense—as ignorance of what a refutation even is—later interpreters took him to be describing a particular fallacy of misrepresentation or irrelevance in refutation. So A's refutation of B is ignorant to the extent that it does not capture what is wrong with B's case, but with something else that seems like B's case. This observation, however wrong, is not unwelcome, because the resultant account of "a man of straw of his owne making" is a very serious kind of argument problem. Indeed, as we're arguing here, it is a problem that takes us to the very core of what argument in dialogue is.

We have outlined our reasons for this. We've mentioned above that the everyday language of the argument is *igiturgid*: it is self-referentially evaluative.

To say "therefore," "thus," or our favorite (of course) "ergo" is not only to perform actions like shooting a hockey puck or sautéing garlic. Unlike these activities, however, to infer is to do something *and* self-evaluate in the process. To modify a phrase from an eminent logician, "to infer that p is to give yourself a little high five," a kind of epistemic pat-on-the-back for a job well done (Nolt 1997: 114). You cannot help but feel the elation of illation. But we have also argued that argument is a meta-phenomenon in another crucial way, the contrastive nature of reasons. Our reasons are good ones, and our illations are worthy of elation, to the extent that they are superior to appropriate contrasts. When they are not, our illations occasion consternation. Both of these characteristics of argument have shaped our discussion of straw manning.

Our main conclusion from these observations is that argument itself, any argument, is essentially meta-argumentative. It is a reflection upon itself and a reflection or an evaluation of the standing alternatives. So, even if you're not straw manning someone right now, the fact of endorsing an argument for p rather than q includes an inchoate representation, even if only to judge the case for some q not worth thinking about. That's how straw manning works, but it's also the key to how reasons work. There is a normative lesson in this. When A reasons their way to p, and endorses p as true, A is also dialectically responsible for the relevant contrasts. A's argument for p therefore contains a representation of the rejected case for some alternative q. So "therefore" at the end of it all not only announces that the reasoning is finished but tells that the reasoning was normatively successful in arriving at the conclusion and eliminating the contrasting alternatives.

While our case for this so far has focused on dialectical requirements, there are doxastic ones as well. Many have observed that it is not psychologically possible for A to hold p while at the same time recognizing that the case for q is stronger. This is the thesis of doxastic involuntarism. I don't have any choice but to applaud myself for my well-formed beliefs. If I don't do so, then I applaud myself for my well-formed beliefs about the ill-formed nature of my other beliefs. Even in thinking that I'm mistaken, I can't escape high-fiving myself for figuring out that I'm mistaken. Having a view isn't just assessing what it is about the fact taken as true to be true but it's also about endorsing that assent. The doxastic case for contrast goes further. Just as we see ourselves as doxastically sound, we see our disagreeing fellows as doxastically deficient. What is important, however, for the case we're making here is that our diagnosis of them takes cognizance of what we take (maybe mistakenly) to be their reasoning for those other contrary and contrasting views, perhaps q. Trudy Govier, in her famous defense of argument's minimal adversariality (1999), noticed that to believe that p requires one to see others who think the contrary to be mistaken. Now we are in a position to see what *mistaken* means in this connection. That someone who disagrees us has *contrary, and therefore false*, beliefs is not the end of the story for us. We have a diagnostic perspective on their being mistaken such that *the way they arrived at their attitude* is mistaken or incomplete; they have weighed the relevant reasons badly, and their doing so occasions a responsibility for me to coordinate with my reasons. The long and the short of it is that the disagreement is more robust than holding competing propositions; it's holding conflicting reasons.

We think this meta-argumentative approach has some distinct advantages. First, let's return to the history of informal logic. Fallacy theory more specifically has its origin in, as we've said, a misreading of Aristotle's *Sophistical Refutations*. Writers of guide books to informal reasoning took Aristotle to be offering a broadly normative account of good dialectical reasoning, when it turned out that Aristotle was discussing something quite different. Naturally, this doesn't mean Aristotle cannot be read in this normative way. If anything, the persistent misreading of Aristotle has shown how fruitful that program was. (Perhaps a case in itself for a salutary straw or iron man, depending on whether you think the misinterpretation was an improvement or debasement, right?) But it has its limits. One of its limits was a narrow focus on free-standing dialectical schemes. These schemes are, as we noted in Chapter 1, often poorly designed specimens of arguments in real life. While there are perfectly good reasons to rely on fake examples, it is true that they occasion meta-argumentative fallacies of their own. One of these, arising from the proliferation of fallacy names, is the now familiar Harry Potter problem. Another of these meta-argumentative fallacies is what we call (noting the irony by the way) the Eager-Beaver problem. The Eager-Beaver problem is a consequence of almost any normative meta-language. Eager new learners are going to make the mistake of overusing the meta-language in their first-order negotiations, and they will apply it to every circumstance. But, as we've discussed in Chapter 5, there is one crucial way that the sports foul analogy for fallacies is illuminating. There are certain kinds of regulative problems that occur only when we try to regulate ourselves. They arise only because of the regulation. They are meta-problems of regulation. So, for example, tripping, in soccer, is a first-order violation of the rules. The existence of the rule against tripping and the consequent penalty for it gave rise to the problem of people pretending to be tripped, or diving. Unsurprisingly, the remedy for this is another rule, now about *diving*. And this is a kind of second-order violation, since it appeals to the rules but does so inappropriately.

Let's pause a moment with the fallacy-foul analogy. This was a metaphor we've been fond of on occasion ourselves. We're not alone. One thing that made the rounds in philosophy a few years back was an image of an American football referee hand signals repurposed to the task of calling fallacies during conference discussions. All the usual suspects had a sign, as well as ones that happen in philosophy in particular. So, statements such *Unilluminating Appeal to Skepticism, Your Claims Are Inconsistent, Naturalistic Fallacy*, and *Please Let the Speaker Finish Before You Object* all had signals. It was funny, and it was shared widely among the philosophers who like well-run argument. The idea is that we all pine for a neutral referee with a whistle and a striped shirt who can sum up someone's dialectical failure with a semi-descriptive hand gesture. Hands pushing outward as if to knock something down would be our proposal for the foul of straw manning. As though all one's opponents are *Push Overs*. That's entertaining so far as that goes. But it doesn't go very far. One reason it doesn't go very far is that it doesn't really capture how fallacies inhabit arguments and how we detect and correct them. The referees are not, as we've said, external to the arguments. With argument, we are the participants and the referees. That is, we don't just argue but we argue about how we are to be arguing—and there's nobody else there to sort it out for us. It's just us. And so when we argue together, we have terms like *point of order*. And we point out when someone's challenges are more filler and not particularly relevant—we call it *quibbling*. We wear two hats as we argue—the one colored for our team (the view we think is right) and a striped one as a referee. And being a member of the arguing community means that you learn to wear both hats, and we learn to co-referee. And we switch out our caps as we go along. For a while two arguers may just wear their red and green hats respectively, but one may need to pop on the striped hat to note that something went awry, and not just from the perspective of their view but from the perspective of the exchange. Our hope is that if we do it right, we keep it all straight—and especially when we put on the ref's cap. But that's precisely the problem: when others put on the ref's cap, it is way too easy to think they're really still playing for their own team.

Strangely (or not, at this point) these problems suggest yet another meta-problem of fallacy theory. The reliance on fake examples is reliance on so many straw men. As Maurice Finocchiaro (1981/2005) has noted, this carries with it a bitter irony: a key tenet of any sound approach to critical thinking or informal logic is that one shouldn't commit the very sin one is railing against. So, populating your scholarly work (ahem), texts, tests, quizzes, examples, and exercise sets with ridiculously weak made-up arguments is to be guilty of straw manning and therefore hypocrisy. This will have especial bite if our current readers of the examples see themselves implicated in them. That is, in a moment of reflection, we think: here we are, writing a book about the straw man fallacy, and we go out finding weak arguments or making them up, just to make a big deal about how bad they are. That's ironic, right? We think (as we discussed in Chapters 1 and 4) this point about the irony misses one important point about straw manning while it hits the bullseye on a fundamental problem anyone who attempts to write or teach about fallacies faces: *How do you represent reasoning that's gone off the rails that at the same time is appealing enough to trick us into thinking it's good?* At first glance, this shouldn't be a problem at all, because we reason badly all of the time. The problem, however, is that we need to make the errors plain to others. And this objection has it that the tiny examples of scheme, however convenient or justified for limited pedagogical or theoretical purposes, do not do this.

Dialogical approaches, such as Walton's or that from pragma-dialectics, seem to point to a way forward. In the first place, they bring some realism with them, and thus obviate the problem of perpetual straw manning examples of fallacy pedagogy and theorizing. In particular, they focus on fallacious reasoning with context and real believers. Second, they model the basic dialectical feature of argument where one reasoner interacts with another. A key problem, however, is that in modeling argument in this external way, dialectical approaches to argument sacrifice too much explanatory power. For one, as we've seen, it's hard to explain the persistence and effectiveness of certain kinds of fallacies, the straw man among them (but also many others, more on those shortly) when they're viewed second personally. You cannot, without a lot of subtlety, straw man someone to their face. But people straw man each other in significant ways all of the time that cannot be captured by a dyadic dialogue. We think, by contrast, a richer and more fecund program is to be found by looking to solve the dialectical issues by turning argument inward. The dialectical feature of argumentation is the result of an argument exchange which is distinct from reasoning. We think *reasoning* is dialectical, or more deeply contrastive, from the beginning. Now certainly, a dialogical model does capture the fact that our reasons are essentially other-regarding. But this is a feature of the reasons as reasons, and as we make this explicit with each other, we must engage in meta-argument. And this is where unique kinds of errors are possible.

7.3 Whataboutism and Bothsiderism

The meta-argumentative program alerts to new, as-of-yet under- or untheorized argumentative pathologies. We've already discussed some of these—the Harry Potter fallacy, the Fallacy Fallacy, the Eager Beaver Fallacy, Burning Man, strongly meta-argumentative straw man, and others. In addition to these, we can add two more significant ones: *Whataboutism* and *Bothsiderism*. Briefly, whataboutism is a technique of diversion, where one responds to a charge by raising a countercharge which is purportedly equal or worse in seriousness. So, for example, Democrats complain that Trump seems to be encouraging violence among his supporters, and Trump supporters respond with "What about left-wing violence?" An example may make this point clearer. Consider the following case.

7.3.1 Biden Train

In the week prior to the 2020 US presidential election, candidate Joe Biden was scheduled to visit Austin, Texas. Texas was considered a swing state at this point in the election, and Biden's visit was welcomed by Democrats, but resisted by Republicans. In fact, a "Trump Train" of privately owned (mostly) pickup trucks blocked Austin's incoming roads to prevent buses for the Biden campaign from entering Austin. Things escalated on the highway, with multiple pickup trucks trying to run the Biden campaign bus off the road and damaging the vehicle of one of Biden's staffers. Videos were circulated. Trump encouraged them and their actions, tweeting, "I LOVE TEXAS" with a video of the incident.[1] And in response to criticism that this behavior from Republicans and the president risks significant violence as the election looms, Texas State Republican Party chairman, Allen West, replied by invoking broader violence between Trump supporters and those supporting the Black Lives Matter movement:

> Three Trump supporters have been executed, one in Portland, one in Denver, and one in Milwaukee. A leftist mob attempted to storm the house of the McCloskeys, threatened to burn their house down, rape Ms. McCloskey, and kill them both. Where is the liberal corporate media's concern about that real violence? ... It is more fake news and propaganda.[2]

West's response was to take a question about the incident with the Biden bus as an "attempt to portray conservatives as violent radicals, even though it is leftists from Antifa and BLM who have been assaulting, robbing, and looting fellow citizens and their property." So the reply was not to address the charge or answer a question but to *ask about other crimes*. In point of fact, this particular version of *whataboutism* is posited on a misrepresentation of the facts, as the "executions" were results of street violence prompted by the so-called "conservatives," and the McCloskey incident was one wherein the couple came out of their home armed to confront BLM protesters headed elsewhere.

Regardless of the misrepresentations, West's reply is to re-direct attention, and in so doing not only avoid answering a critical question of what to do about violent acts directed toward candidates for public office that are encouraged by the competing candidate, but then uses this as an occasion to make the case against those who'd asked the original question.

So with the *whatabout* argument, the speaker replies to a target with a widened set of considerations. Some may be relevant to the matter, but the problem is that if asked about a particular incident (the Biden bus case on the road outside Austin), talking about another set of incidents (riots in Portland and a St. Louis couple waving guns at protesters and then being surprised that the protesters didn't like it) evades the original question. And the implication is that there are more important things to talk about—other issues need evaluation. And so it makes a case against the judgment of those who'd posed the original query. It is a meta-argumentative move in the sense that it is (a) about the *argument at hand*—that *this* is not as important as *that* issue, and (b) it is about the *arguers* in the exchange—that there is a problem of inappropriate emphasis and likely an instance of bias in the background. And it is clear that one can legitimately make such a meta-argumentative move. There are many instances where we are distracted by a small thing when bigger things must be discussed. But what this does is, like with the straw man, use the meta-argumentative case to prematurely close a critical dialogue.

Certainly, the *whatabout* charge strongly recalls the *tu quoque* hypocrisy charge, which criticizes a claim proposed by a speaker on the basis of the fact that the speaker otherwise does not live in ways consistent with it. However, it also suggests the strongly contrastive notion that arguments for some conclusion p fail because they do not take cognizance of the relevant contrasts. So, for instance, A may morally object to p, but there are other cases of equal or greater importance, so A ought to object to those as well. Given that A has not objected to those, their concern for the case of p is unjustified. Or alternatively, perhaps the contrasts show that the case for p's being objectionable is not as strong, and so A should be shocked and appalled by q. Along somewhat similar lines, Aspetitia (2020) also reads *whatabout* arguments dialectically, holding that they arise from the suspicion that an interlocutor is not expressing their real views. In this sense, they have strong resonance with the representational form of the straw man.

Bothsiderism is another interesting case of a meta-argumentative fallacy. To put it very roughly, Bothsiderism is when one concludes that because there is live controversy between two (or more) other arguers, the matter is unknowable, unresolvable, or that making a decision on it would be hasty. So bothsiderist arguments have the opposite function of what we'd seen with straw man arguments—they keep discussions open that should be counted as resolved and thereby closed. Bothsiderism we think is another perfect encapsulation of the meta-argumentative problem because it regards, to put this bluntly and somewhat hilariously, the taking of contrastive reason's contrastive reasons. It is a kind of meta-straw man. As a historical matter, Whately (1826) saw it as a broader objective of what we've called weak manning and iron manning. One distorts the dialogue for the sake of a discussion that provides further meta-argumentative evidence about the issue. In short, if the discussion gets dragged out, the participants aggregate evidence that the issue is not so easily resolved. So we must suspend judgment. It is, we think with some issues, absolutely insidious and pervasive.

Like whataboutism, it is not hard to find examples of bothsiderism in real life. In the wake of a violent confrontation between racist white nationalists and protesters in Charlottesville, Virginia, in August 2017, President Trump famously remarked that there are good people "on both sides" of the confrontation.[3] As if to say the disagreement between white nationalists (including Nazis and members of the Ku Klux Klan) and their opponents is an honest disagreement between citizens, akin to a serious and well-researched discussion over how much to invest in green energy. The warped sense of balance extends to scientific questions as well since bothsiderism promotes a kind of conservatively amenable skepticism. Consider the following.

7.3.2 Barrett on Warming

In the Senate hearings for confirming Amy Coney Barrett to the US Supreme Court, then senator (now vice president) Kamala Harris posed a set of questions which elicited an interesting answer:

Senator Harris:
And do you accept that COVID-19 is infectious?

Amy Coney Barrett:
I think, yes. I do accept that COVID-19 is infectious. That's something of which I feel like we could say you take judicial notice of. It's an obvious fact, yes.

Senator Harris:
Do you accept that smoking causes cancer?

Amy Coney Barrett:
I'm not sure exactly where you're going with this, but the notice that smoking causes—

Senator Harris:
[crosstalk] Question is what it is. You can answer it if you believe it. Yes or no.

Amy Coney Barrett:
Senator Harris, yes. Every package of cigarettes warns that smoking causes cancer.

Senator Harris:
And do you believe that climate change is happening and it's threatening the air we breathe and the water we drink?

Amy Coney Barrett:
Senator again, I was wondering where you were going with that. You have asked me a series of questions that are completely uncontroversial, like whether COVID-19 is infectious, whether smoking causes cancer, and then trying to analogize that to eliciting an opinion from me that is on a very contentious matter of public debate. And I will not do that. It will not express a view on a matter of public policy, especially one that is politically controversial because that's inconsistent with the judicial role, as I have explained.[4]

Barrett here offers two reasons for suspending judgment on climate change. First, as a pragmatic and professional matter, she may have a case, and so finds it imprudent to express an opinion when she might have to judge the matter. This is not all that convincing, for we might note the same could be true of smoking; that litigation is certainly not over. The second reason is the meta-argumentative one. She notes that the "very contentious matter of public debate" is a matter of significance, and as a consequence she thinks she should have no settled opinion on the matter. In other words, that there are arguings about p is evidence about the extent to which p is true or settled. And since it is evidence that there is conflicted evidence, we should suspend judgment. This was advice the skeptics used, in the classic forms of the argument from disagreement, to give to bring about suspending judgment.

Like straw manning, bothsiderism closely mimics dialectical argument norms. Good reasoners certainly ought to consider both sides of a debate, even when the arguments for one point of view appear to be weak. After all, awareness of differing or opposed opinions is one hallmark of a person who is authentically and competently critically engaged on an issue. Beyond that, it's generally a good thing that we are on our guard against journalism that is *one sided* and for people who are *biased*. We are thus praised for being *even-handed* or *balanced* in adjudicating facts. That such virtuous argument practices are evidently exploitable is a troubling fact.

Despite the *prima facie* goods to bothsiderist approaches in maintaining open-inquiry norms, the epistemic risks of this meta-argumentative move are legion. One risk is that it encourages arguers to hold their ground long after they've exhausted their chances of success. Just staying around to argue is meta-argumentative evidence to certain audiences. Bothsiderism also conveys a tactical advantage to weak views. For Millian reasons, people confer great value on dissenting views, however unpopular. Note again the outsize presence of Climate Change Skeptics in the public mind.

We can even tie Bothsiderism to weak manning and, in turn, to iron manning. To show that a case is not settled, it is sufficient to show the inattentive or otherwise biased listener that the case for one side isn't perfect, because *doubts remain*. So the epistemically virtuous thing to do is suspend judgment. However, we can see how this weak mans a strong case, as it foregrounds problems for the strong case, instead of the overall balance of reasons. The case of iron manning with bothsiderism is very similar. One needs only to exaggerate the significance of objections to draw the same skeptical conclusion—the matter remains in dispute and nonexperts are going to have to suspend judgment, or draw the conclusion that both sides have some merit. Barrett's attitude about climate change above is certainly exemplary of this attitude and tactic, as, by any fair reckoning, there really are not two sides to the climate change issue. We have argued that the straw man is fallacious on account of the meta-argumentative closing function. And something odd is at work in bothsiderism that is a mirror-image of what we'd seen in straw man fallacies. That is, straw man arguments close arguments too soon, and bothsiderism works to hold dialogues open beyond when appropriate. In the end, bothsiderism also points up another crucial aspect of meta-argument. Many of our views, even our most developed views, are views *about* arguments, about disputes. We have a sense of what the sides are and who the major players are and whether they are still open or closed issues.

Now to be clear, our aim here is not to give a full account of whataboutism or bothsiderism. We'll save that for another time. But we will close this discussion by noting that bothsiderism underscores another salient aspect of an account of argument badness. It is naturally a kind of truism to say that a fallacy is an argument that *seems* to be a good one but isn't. Our first puzzle was with effectiveness, because there is no point in talking about fallacies that are not effective. And the problem is that fallacy examples oftentimes suffer from this as a paradoxical consequence of their placement in a text on fallacies. They are by notoriously ineffective. But that really should not matter, because their function as textbook examples is not to trick anyone. While they serve, when well-written, to bring our attention to certain kinds of failings, they can't re-create the experience of being duped and they struggle to represent what is really at the core of our account of the fallaciousness of the straw man. The very point of their presentation was to break their own spells so that we are not moved by them or their ilk. They can't represent the diachronic or downstream effects of the straw man and related meta-argumentative strategies. One reason for this is that, for pedagogical reasons, they are often *conclusive*. The key with bothsiderism is that it undercuts the conclusivity of cases by recasting the state of dialectical play. What, among those best informed, is settled is presented to the uninformed as not. This is done, again, not by way of reasoning about the matter but reasoning about the reasoning on the matter. But that meta-reasoning is not reasoning in the reasons given on the first order beyond the simple fact that there happen to be dissent and dissenters. Not once, for example, does Amy Coney Barrett reference data or evidence about climate change—she references the fact that there is controversy about the issue. And this is sufficient for her to say the matter is not settled. In fact, as we see here, the bothsiderist program can be performed without, really, any knowledge of any of the first-order evidence—one seems to need only to point to there being someone who disagrees with someone else. Their epistemic standing or their grounds for doing so is beside the point in the bothsiderist presentation. All that is required is for there to be a disagreement. And, as we've observed before, one of the curious things about the meta-argumentative fallacies is how they have a tendency to have little connection to the first-order questions they bear on. In fact, they are studied evasions of them.

7.4 The Owl of Minerva Problem

The phenomenon of what we've called strongly meta-argumentative fallacies is a curious and ironic one. So from "Ann Coulter on *Ad Hominem*," "Goldberg's Strict Constructionists," and "Obama's Demagoguery," we see a troubling phenomenon—that the vocabulary of fallacies makes totally new meta-argumentative fallacies possible. You can't commit the Harry Potter fallacy or the fallacy fallacy fallacy without all the apparatus of fallacy detection and correction assembled. And before you say *Well what about <ahem> all those first-order fallacies, don't they matter?*, we will note that the challenges we are talking about here arise when we start trying to correct those first-order fallacies. So, "Coulter on *Ad Hominem*" works the way it does because there are, we admit, plenty of cases of character assassination in political arguments. Both sides do it plenty (we desperately hope you see what we're doing here), so the corrections are there because we correct first-order arguments with the meta-language. And we do make those corrections; we, fallible creatures that we are, mess things up on that level, too. That's just the kind of beings we are.

The Owl of Minerva flies only at dusk. That's a poetic way of saying that *wisdom arrives only in hindsight.* Hegel invoked the image to capture the thought that philosophy arrives too late to the game with its tools to explain what's happened, painting gray on gray with fine distinctions and so on. Our reflection on ourselves is retrospective—there must be something upon which to reflect, and that's the past, what we've done. And as we come to know things about ourselves, reflective creatures that we are, we change in light of that knowledge. But that makes us moving targets for that reflection—if our reflection changes who we are as we live in light of it and reflection bears on what's been done, then we have a rough picture of the problem. And it is revealed in fallacy theory with the hopeful face of the program of critical thinking (it's the story we tell the deans and our colleagues about how good our classes are for students and society generally). These changes brought by reflection are salutary—we detect fallacies and so improve our thinking. But there's bad news, too, and it's that our fallibility that was in need of correction on the first level moves up with us to that next level. And new illusions and new errors follow. That's the secret bad news of it all, and it's not a thing to pitch to the university donors that these reflections will make new (but *very interesting!*) pathologies of reason possible. We, again, don't think there is a solution to this problem, as it comes by way of our capacity to detect errors in reasoning. Only being made gods would cure us of it, that is, only if we were made infallible. But, of course, that would have also eliminated the need for the tools by which we commit these meta-argumentative errors to begin with. Gods don't need critical thinking tools—the tools are tools only for fallible and reflective creatures. What we need, then, is a set of skills for managing this problem, not solving it. Unfortunately, there are not many good models yet for this plan.

Metaphilosophy is a site of this kind of problem. In fact, the pull of metaphilosophy is quite strong and perfectly reasonable. Most every philosophical dispute has the potential to go meta- at any time. So philosophers X and Y may disagree about free will, and after they've shared their reasons, one will invariably diagnose the others' errors not just in their reasons in the strict sense but in their reasons in the broad sense. They come from a tradition that takes intuitions too seriously, or not. They are too scientistic, or are scientifically illiterate. On the one hand, it's a good explanatory tool for why the dispute looks the way it does and why it seems so hard to solve—now it's not just about this philosophical problem but a philosophical problem about how to solve philosophical problems. And we can see now why all this clarity makes resolution harder. In fact, once on this level, the original dispute seems to recede from view. We've both suspected that turning to metaphilosophy is a dodge from actual philosophical argument. As if to say: *Oh, the evidence goes against my view? Well I'm not working within your epistemic paradigm—what do you say to that?*. The painful irony is that we, ourselves, just did the bad thing, too, since to make the point, we had to make the metaphilosophical move up a level.

And notice now a parallel in political argument. The moment that someone starts talking about the tone of political debate these days, we at first find ourselves agreeing. It is lamentable that we are so fractious. But then we start suspecting that it is a way of not articulating any particular policy or political view at all. In the months leading up to the 2020 election in Tennessee, a local candidate's whole commercial message was posited on how he'll "change the tone of politics in Tennessee." First, that's some real nonsense there—no individual changes the tone of a fractious culture, so this is a pretty empty promise even if he really means it and wants to try. But second, we suspect it's a *dodge*—we get no information about what he supports or doesn't. Just that he'll be really polite about it all while we're discussing it. The point, again, is that going meta- can be, as we noted with "Barrett on Warming," a studied avoidance of addressing any issue of substance. Doing the work of thinking things through and having a reason-tested view, being vigilant. Being about tone first and last is a kind of pantomime of that thoughtfulness.

The Owl of Minerva problem isn't, by our lights, solvable. But we think it is manageable. The first site of management is, just like with the Harry Potter problem, to refocus our attention. The reason why the meta-language of logic has the corrective power it does is because there are arguments that work and those that don't. What the meta-language does is give us the tools to explain what's gone right and what's gone wrong, and it allows us to make corrections. So the first bit of management is to keep our attention focused primarily on the issues at hand and the evidence bearing on them—the meta-tools are for that purpose, not their own free standing good. So, in philosophy, we should focus on the first-order issues and go meta- only when we absolutely must, and we should do our best to stop doing metaphilosophy as soon as possible. Metaphilosophy is too many philosophers' habit of climbing into their own navels. And in political discussion, we should focus on the issues and leave the climate questions and tone policing only for times they truly impede first-order discussion.

The irony, again, is that to say all this we must have climbed the ladder which we insist we must climb only infrequently, if ever. That's not a self-refutation, but it's not exactly a comfortable consistency, either. And note that even our advice for managing the Owl of Minerva problem has itself an Owl of Minerva problem. Our recommendation not to go meta- in philosophy now can be abused by those who cut off discussion because it hinges on a question of clashes of broader philosophical method. Our recommendation to focus on issues instead of climate can be abused by those who would use this recommendation to silence complaints that the tone of a debate sidelines vulnerable populations impacted by it. It, of course, is not a surprise that our analysis and proposals for managing the Owl of Minerva problem themselves have an Owl of Minerva problem. That's the game once meta-argumentative errors are the salient issue.

And so there's good news and bad news. The good news is that we've begun to assemble some tools to diagnose and explain some meta-argumentative errors. Straw man being one, but it is a cousin to many more. The bad news is that we showed that meta-argumentative errors are strengthened by the very vocabulary we've used to detect them. That's the Owl of Minerva problem for fallacy theory, and we just don't think it is a soluble problem. It is manageable, but even *that* strategy has its own version of the problem plague it. It's all ironic, but by now, we expect that for our readers, it's not surprising.

7.5 Conclusion

It is time to take stock. The straw man fallacy is worth getting right, not just for theoretical purposes but also for the purposes of better critical thinking pedagogy. The fallacy is interesting. It has many forms—representational straw man, selectional weak man, and the hollow man. The fallacy aggregates with what we call the burning man, and it turns out one can even strategically straw man oneself (what we called the self straw man). The argument type is regularly fallacious, but there are argumentatively salutary versions of it. And this shows that the misrepresentation at the core of straw manning does not explain why it is a fallacy. Rather, we've shown, the premature closure of a critical dialogue is what makes straw man fallacious, as the salutary cases of misrepresentation open the dialogues in fruitful ways. Additionally, the straw man as a negative representation of a target has a positive cousin, the iron man. And it has salutary instances where one improves the prospects for the alternatives in shared reasoning, and it has fallacious instances, too, with improvements of views that distort well-run dialogue that otherwise would close on an issue.

We've argued that the straw man fallacy has three puzzles for a proper theory. The effectiveness puzzle was: how could the criticism of a misrepresentation of a view be taken to be a successful critique of the original view? The dialecticality puzzle was: how could criticism of a bad view or unacceptable argument itself be fallacious? And the meta-argumentation puzzle was: how could our reasoning about reasoning (and, in particular, our avoidance of bad arguments) itself yield a bad argument? We think we've answered these puzzles by showing that straw man arguments arise from polyadic dialogues, and so can be presented in *second-* or *third-* person. They move their audiences differently depending on the audiences they address. And if the misrepresentation is presented as the relevant contrast for our own views, the straw man misrepresentation uses our dialectical attunement and meta-argumentative scrutiny against us. The arguments and views presented for critique are, by hypothesis, bad, so we infer that, if they are representative, then we have, on the meta-argumentative level, closed the issue.

Additionally, we've shown that the straw man fallacy arose out of misreadings of Aristotle's *ignoratio elenchi*, but they maintained his core dyadic dialogical model for exchange in the development of the theory. The metaphor of *straw* captured the misrepresentation of one's opponents as weaker and easier to overcome, and by the modern period, this misrepresentational model for the fallacy was its defining feature. This is borne out not only by the textbook accounts of the fallacy but by the going scholarly accounts in the pragmatic, pragma-dialectical, and rhetorical traditions. We've shown that the misrepresentational core of the fallacy as fallacy isn't quite right but it certainly is part of the explanation for how things can go wrong in a critical dialogue.

Finally, we've argued that our approach, invoking onlooking audiences, foregrounding contrastivity of argumentative reasons, and highlighting the meta-argumentative norms in critical dialogue, is a fecund research program in and of itself. We think this polyadic, contrastive, and meta-argumentative approach does a robust job of properly theorizing the straw man fallacy, but we've also shown that there can be useful roles to play beyond. In critical thinking pedagogy, we have the Eager Beaver and Harry Potter problems, and we think identifying these fallacies and redirecting our attention to heuristic meta-argumentative skills is useful. In theorizing argument generally, we have made the case for a theory of the intrinsic adversariality of argument, and this model helps explain so many of the problems that give rise to the fallacies, and it helps us find paths to managing that adversariality with argumentative virtues such as open-mindedness (and knowing its limits). Further, we think our program opens avenues for theorizing meta-argumentative fallacies, starting with the straw man, but extending to bothsiderism and whataboutism. And, in closing, we want to highlight that with this outline, we have articulated what is precarious about argumentation theory: we are arguing about argument, and because we are fallible and reflective creatures, we don't just err when we reason about things but we err when we reason about how we reason about things. And many of those errors are unique to that meta-level of argument. The straw man, again, can be committed only by reasoning about reasoning, and it can be made into what we've called a *strongly* meta-argumentative fallacy by making the fallacies attributed explicit. Our hope with theorizing the fallacies properly was that we make it so that they no longer cast their spells over us. But with these new skills of meta-argument that we acquire, new otherwise unthinkable spells and illusions await. That's the Owl of Minerva problem, and it's progress of a sort. It's something that we, as creatures resolved to be vigilant amid our fallible reasoning, take as a bracing truth.

Notes

Chapter 1

1   It is worth noting that Hansen (2002) has corrected this thesis that the standard view was widely held.
2   Versions of this problem arise regularly in axiology. This is why Aristotle, even after he holds that ethical norms must be universal in a sense, requires that judgments of actions must have sensitivity to the particulars (NE 1098a.33 and see Leibowitz 2013). In the twentieth century, the issues arose in ethics with the challenge of particularism as in Prichard (1912) and in epistemology with the generality problem for reliabilism Conee and Feldman (1998).
3   Here is an exercise we've had success with. Rather than have the students go on the hunt for fallacies (a practice we actually deplore), have them commit them on purpose. It is remarkable how easily the slippery slope slides off their pen.
4   A similar line of thought goes for examples of deductive arguments.
5   Director and Second World War veteran Samuel Fuller reportedly said of war movies: "The only way to bring the real experience of war to a movie audience is by firing a machine gun above their heads during the screening."
6   See Fogelin and Duggan (1987: 256) for this model, and Wittgenstein (1953: S67) for the reference to "family resemblances."
7   See van Laar (2008) and Anderson, Aikin, and Casey (2012) for developments of the notion that fallacy-charges play dialectical roles in the clarification of views. Bermejo-Luque (2010) argues that such concepts allow a more clear recursive dialogical form for arguments. Further, Weinstock, Neuman, and Tabak (2004) have evidence that shows that learning the vocabulary of fallacy theory and norms of argumentation yields a sensitivity to instances of fallacies and makes one less likely to be moved by them.
8   For a discussion of this very example, see Schechter (2019).
9   Here's how the fallacy fallacy fallacy would run. Imagine an argumentative context with asymmetric burden of proof, a courtroom perhaps. So, if the prosecution's arguments fail, we revert to holding the defendant not-guilty and reject the guilty verdict. Now, imagine the prosecution's case is entirely fallacious arguments. So, the defense can argue that since the prosecution's arguments are fallacious we should reject the guilty verdict. Now imagine the prosecution charging the defense with

committing the *fallacy fallacy*. Just because the case for guilt is bad, it doesn't follow that they are innocent. But to make this move in this context is to inappropriately apply the concept of the fallacy fallacy, because the burden of proof is on the prosecution. So, the prosecution, in making this charge of fallacy fallacy against the defense case against them, has committed the *fallacy fallacy fallacy*.

10 Others who have noted this connection are Paul (1984); Johnson (2000: 243); Cohen (2004); and Boudry, Paglieri, and Pigliucci (2015).
11 Aikin and Talisse have taken on the project of adding to the fallacy-theoretic program a set of methods of repairing argument and argument cultures (see 2019, 2020).
12 In this, we agree with Hundleby's proposals that fallacy theory and fallacy instruction need to be taught alongside the broader research in informal logic, particularly that of argument repair (2010: 299).
13 See Aikin (2011a) for models of argumentative escalation and accounts of its mitigation.

Chapter 2

1 Locke did not invent the term "ad hominem." Hamblin (1970: 161–2) surmises the he retrieved it from medieval translations of Aristotle.
2 It has been noted, however, that Hamblin leaves out all of rhetorical history in his discussion of the standard account. See Michel Dufour (2019) and Hansen (2002).
3 For helpful descriptions, see Duncombe and Dutilh Novaes (2016) and Krabbe (2012).
4 See Casey (2012).
5 Peter of Spain, for example. See *Summary of Logic* (2014: 385).
6 See Casey (2020b).
7 This is somewhat akin to Aristotle's remarking that *ignoratio elenchi* is both a general failure—every fallacy is a matter of ignorance of the refutation—and a particular one, where *ignoratio elenchi* is a specific kind of failure.
8 Cf. Baronett *Logic* second edition (2013).

Chapter 3

1 See Walton (1996: 125) for this analysis.
2 This feature of misrepresenting the opposition's quality of case is distinct from other dialectical misrepresentations—for example, denying that one previously agreed to some claim, allowing evidence from a source but then rejecting it later, and that of feigning doubt in order to increase the opposition's burden of proof.

3 Of course, the speaker should be able to respond effectively to even the trifling objections if prompted, but this is not a requirement in place at all times.
4 It should be noted that there must be an additional social factor for the selectional straw man fallacy—there must not only be better cases (C's perhaps) than B's for A to address, but C's case must be one that if A is at all a responsible researcher of their general opposition's views and arguments, they would have come upon C's case and recognized it as such.
5 Analogize this strategy to that of gerrymandering evidence, or "nut picking." Gerrymandering evidence is the act of selecting only evidence that supports one's preferred thesis or undercuts the opposition's. In weak manning, one gerrymanders one's evidence of the opposition's views by selecting the worst of the opposition as representative.
6 Of course, A can still misrepresent her *selected* straw man. This will transform a selectional version of the fallacy into a representational form.
7 For this reason, Walton argues that the straw man fallacy is a special form of *secundum quid* (1996: 122).
8 The conditions for this are clearly slippery, but at least showing that A's argument does not touch arguments posed by C and D is a start. Additionally, showing that these neglected arguments are from superior sources is an option—for example, the more recent, those in better journals or more reputable media sources, from widely recognized authorities as opposed to less recognized.
9 Mill writes, "He who knows only his own side of the case knows little of that" (1991: 42). It should be noted that the intuitive force of this Millian principle partially explains the success and prevalence of the selection form of the straw man fallacy: in order to convince, one must give one's audience the impression that they have understood adequately the opposition.
10 https://www.nationalreview.com/2017/03/donald-trump-media-apocalyptic-rhetoric-unhealthy-unsustainable/
11 https://www.washingtonpost.com/politics/2018/10/10/eric-holder-when-they-go-low-we-kick-them-thats-what-this-new-democratic-party-is-about/?noredirect=on
12 Video of Hannity's show is available at the following address, start at 6:19 https://video.foxnews.com/v/584/885499001/?playlist_id=5531425782001#sp=show-clips
13 https://www.huffpost.com/entry/trevor-noah-fox-news-phony-victims-unit-eric-holder_n_5bc0c083e4b0bd9ed559b520
14 https://youtu.be/NoXGV4Vw-VA
15 https://www.zerohedge.com/news/2017-02-22/msnbc-anchor-admits-our-job-control-exactly-what-people-think
16 https://www.breitbart.com/clips/2017/02/22/brzezinski-controlling-what-people-think-is-our-job/
17 https://twitter.com/morningmika/status/834463604531404800?s=20

18 See Ackerman and Fishkin (2004: 5ff.) for a quick survey of the relevant findings in the recent literature. The classic book-length study of public ignorance is Delli Carpini and Keeter (1996). See also Converse (1964) and Somin (1998).
19 See the study by the PIPA organization, "Misperceptions, the Media, and the Iraq War," http://www.pipa.org/OnlineReports/Iraq/Media_10_02_03_Report.pdf
20 Hollow manning has only infrequently been recognized as a possibility for straw man fallacies. Frans van Eemeren, Rob Grootendorst, and Francisca Snoeck Henkemans present two forms of the straw man. One is along the representational lines and the other along hollow man lines: "There are two different ways of attacking a standpoint that is not really the one presented by the opponent. The original standpoint can be misrepresented, or a fictitious standpoint can be attributed to the opponent" (2002: 117). Johnson and Blair's definition of the straw man leaves open room for the possibility of hollow man, namely that an arguer attributes to another some position Q, but the other arguer's position is actually R (2006: 94). Whether Q is a distortion of R or something entirely different is not explicit in their model. See also Copi, Cohen, and Flage (2007: 445) and Rudinow and Barry (2008: 325).
21 Cf. Sunstein's (2009) work showing that overtly addressing rumors actually backfires in cases where the audience has accepted the rumor as likely true. "Corrections of false impressions can be futile; they can also actually strengthen those impressions" (46).
22 People were asked to respond to various civic proposals, some framed by straw men of the opposition, others without. The variations of interest were affected by changing the geography of where the civic proposals would be put into effect (near=higher interest, far away=lower interest). We discuss this further in Chapter 7.
23 Marianna Bergamaschi Ganapini (forthcoming) has argued that there is an additional signaling function to sharing fake news about one's political opposition—instead of giving new information (which is risky, since it's fake news), it stakes a claim to and reinforces to a particular group identity.

Chapter 4

1 A version of this insight about challenges has been made by Van Laar (2008).
2 https://www.theatlantic.com/politics/archive/2010/09/what-cornyn-didnt-say-about-gays/62364/
3 https://www.pbs.org/video/defunding-the-police-1591649572/
4 See Stevens (2020) and Siegel (1995) for moral cases for charitable inclusion.
5 See Anderson et al. (2013).
6 This notion of dialectical virtue is in the same spirit as the deliberative-epistemic virtues discussed in Aikin (2008), Aikin and Clanton (2010), Ferkany and Whyte (2011), DeBruin (2013), and Scott (2014).

7   See Aikin (2008) for norms bearing on the "sliding scale" of explicitness and clarity to expect of views.
8   No dialectical move without a Latin name is worth making. Sometimes it requires finding the name; other times it requires inventing it.
9   See Aikin (2008); Hare (2003); Kitcher (1982).

Chapter 5

1   https://twitter.com/realDonaldTrump/status/1284282946850086912?s=20
2   https://abcnews.go.com/Politics/fact-check-biden-trump-rival-defund-police/story?id=72554629
3   A compilation of the Dixon Tweet, the Fox Tweet, and the comments is available at https://www.facebook.com/TheBenjaminDixonShow/photos/a.1108591322675973/1538583016343466/
4   http://spectator.org/archives/2013/09/04/obamas-half-assad-war
5   John Dickerson's *Slate* overview of the options makes for sobering reading, as it's clear there are no good routes for any long term policy, given the complications in the region. https://slate.com/news-and-politics/2013/09/obama-congress-and-syria-arguments-for-and-against-the-bombing-resolution.html
6   For the transcript of the first debate, seehttps://www.usatoday.com/story/news/politics/elections/2020/09/30/presidential-debate-read-full-transcript-first-debate/3587462001/
7   Stevens (2021) also argues that an epistemic asymmetry between interlocutors is crucial to the effectiveness of second-personal straw manning. In such a case, one arguer may have a weaker grasp of their reasons than their interlocutor with their criticisms and so the arguer will be more likely to accept a distorted negative assessment.
8   Cynthia Stark rightly notes that gaslighting in public has a broader form of "psychological oppression," wherein not only the target for gaslighting has the picture of themselves presented to them but all those who identify with the target are so negatively depicted (2019: 231). Andrew Spear also observes that such self-conceptions require regular updatings for the confabulations to be stable (2020: 225).

Chapter 6

1   For an account of the distinction between first-order and meta-argumentation, see Finocchiaro (2014), Van den Hoven (2015), Hovhannisyan and Djidjian (2017).
2   https://archives.infowars.com/tv-poll-71-of-liberals-dont-want-peace-with-north-korea-because-trump-would-take-credit/

3   http://www.whatisitliketobeaphilosopher.com/#/bryan-van-norden/
4   Daniel Cohen has termed this the Principle of Meta-Rationality (PMR): "Part of reasoning rationally is reasoning about rationality" (2001: 78).
5   Tesfaye, Sophia. 2015. "Ben Carson is just this Vile." *Salon.com*. Wednesday, October 7. http://www.salon.com/2015/10/07/ben_carson_is_just_this_vile_5_repugnant_statement_hes_made_since_the_oregon_mass_shooting/
6   Prager, Dennis. 2015. "The Right-Left Divide on Gun Control." *National Review Online*. October 6. http://www.nationalreview.com/right-left-divide-on-gun-control
7   https://www.nationalreview.com/2011/07/demagogic-style-victor-davis-hanson/
8   https://townhall.com/columnists/jonahgoldberg/2010/08/11/constitutional-amendments-and-citizenship-rights-n1361264
9   https://www.politico.com/story/2010/08/graham-14th-amendment-outdated-040635
10 https://twitter.com/realDonaldTrump/status/1052213711295930368?s=20
11 https://apnews.com/article/ff81da8870fa434c90214799813ba5bc
12 https://www.thedailybeast.com/trump-believed-his-stormy-daniels-horseface-tweet-was-politically-brilliant
13 See Aikin and Talisse (2018) for an overview of this literature. The implications of this and similar phenomena of indiscernibility for political discussion are drawn out by Aikin (2012); Lambert-Beatty (2009); LaMarre, Landreville, and Beam (2009).
14 http://linguafranca.mirror.theinfo.org/9605/sokal.html
15 https://www.chronicle.com/article/sokal-squared-is-huge-publishing-hoax-hilarious-and-delightful-or-an-ugly-example-of-dishonesty-and-bad-faith/
16 https://www.nationalreview.com/2018/06/dog-park-study-rape-culture-portland-ungendering-research-initiative/
17 See *The Atlantic*: https://www.theatlantic.com/politics/archive/2016/09/trump-makes-his-case-in-pittsburgh/501335/; *Real Clear Politics*: https://www.realclearpolitics.com/articles/2017/04/23/taking_trump_seriously_and_literally_133682.html; *The CNBC*: https://www.cnbc.com/2016/11/09/peter-thiel-perfectly-summed-up-donald-trump-in-one-paragraph.html; *The LA Times*: https://www.latimes.com/opinion/op-ed/la-oe-goldberg-trump-seriously-literally-20161206-story.html
18 https://newrepublic.com/article/139004/ironic-nazis-still-nazis
19 https://www.theatlantic.com/politics/archive/2016/09/trump-makes-his-case-in-pittsburgh/501335/
20 https://www.newstatesman.com/culture/books/2019/08/facebook-vote-leave-how-we-entered-war-against-reality

21. https://www.whitehouse.gov/briefings-statements/remarks-president-trump-signing-ceremony-h-r-266-paycheck-protection-program-health-care-enhancement-act/

Chapter 7

1. https://twitter.com/realDonaldTrump/status/1322700188624932869
2. https://www.texasgop.org/statement-biden-bus-incident/
3. For a complete transcript of the president's remarks: https://www.politifact.com/article/2019/apr/26/context-trumps-very-fine-people-both-sides-remarks/.
4. https://www.rev.com/blog/transcripts/amy-coney-barrett-senate-confirmation-hearing-day-3-transcript

References

Abramson, Kate (2013) "Turning Up the Lights on Gaslighting." *Philosophical Perspectives*. 13. 1–30.

Ackerman, Bruce and James Fishkin (2004) *Deliberation Day*. New Haven: Yale University Press.

Aikin, Scott (2008) "Holding One's Own." *Argumentation*. 22. 571–84.

Aikin, Scott (2011a) "A Defense of War and Sport Metaphors for Argument." *Philosophy and Rhetoric*. 44. 3.

Aikin, Scott (2011b) "The Rhetorical Theory of Argument is Self-Defeating." *Cogency*. 3. 79–92.

Aikin, Scott (2012) "Poe's Law, Group Polarization, and Argumentative Failure in Religious and Political Discourse." *Social Semiotics*. 22. 1–12.

Aikin, Scott (2017) "Fallacy Theory, the Negativity Problem, and Minimal Dialectical Adversariality." *Cogency*. 9:1. 7–19.

Aikin, Scott (2020) "The Owl of Minerva Problem." *Southwest Philosophical Review*. 31:1. 13–22.

Aikin, Scott and Alsip Vollbrecht Lucy (2020) "Argumentative Ethics." In Hugh LaFollette (ed.), *International Encyclopedia of Ethics*. Malden, MA: Wiley-Blackwell, 253–5.

Aikin, Scott and Caleb Clanton (2010) "Developing Deliberative Virtues." *Journal of Applied Philosophy*. 27:4. 409–24.

Aikin, Scott and Robert Talisse (2008) "Modus tonens." *Argumentation*. 22:4. 521–9.

Aikin, Scott and Robert Talisse (2018) *Pragmatism, Pluralism, and the Nature of Philosophy*. New York: Routledge.

Aikin, Scott and Robert Talisse (2019) *Why We Argue*. New York: Routledge.

Aikin, Scott and Robert Talisse (2020) *Political Argument in a Polarized Age*. London: Polity.

Aikin, Scott (forthcoming) "Argumentative Adversariality. Contrastive Reasons, and the Winners-and-Losers Problem." *Topoi*.

Anderson, Ashley, Dominique Brossard, Dietram A. Scheufele, A. Xenos Michael and Peter Ladwig (2013) "The 'Nasty Effect:' Online Incivility and Risk Perceptions of Emerging Technologies." *Journal of Computer-Mediated Communication*. 19:3. 373–87.

Anderson, Colin, Scott Aikin and John Casey (2012) "You'd Sing a Different Tune: Subjunctive *Tu Quoque* Arguments." *Inquiry*. 27. 1.

Aristotle (1999) *Nicomachean Ethics*. Trans Martin Ostwald. Upper Saddle River, NJ: Prentice Hall.

Aristotle (1984) "*On Sophistical Refutations.*" Translated by W.A. Pickard-Cambridge. In Jonathan Barnes (ed.), *The Complete Works of Aristotle.* Vol. 1. Princeton: Princeton University Press, 278–314.

Arnauld, Antoine and Pierre Nicole (1850) *Logic, or, The Art of Thinking, being the Port-Royal Logic.* Translated by Thomas Spencer Baynes. Edinburgh: Sutherland and Knox.

Aspeitia, Barceló (2020) "Whataboutisms and Inconsistency." Argumentation. 34. 433–47. https://doi.org/10.1007/s10503-020-09515-1.

Baronett, Stan (2013) *Logic.* New York: Oxford University Press.

Bassham, Greg, William Irwin, Henry Nardone and James M. Wallace (2002) *Critical Thinking.* Boston: McGraw Hill.

Battaly, Heather (2018) "Can Closed-Mindedness be an Intellectual Virtue?" *Royal Institute of Philosophy Supplement.* 84. 23–45.

Baumann, Peter (2015) *Epistemic Contrastivism. Routledge Encyclopedia of Philosophy.* New York: Routledge.

Baumeister, Roy (2005) *The Cultural Animal: Human Nature, Meaning, and Social Life.* New York: Oxford University Press.

Beardsley, Monroe C. (1950) *Practical Logic.* Englewood Cliffs, N.J.: Prentice Hall, Inc.

Bentham, Jeremy (1824) *Book of Fallacies from Unfinished Papers of Jeremy Bentham.* London: John and H.L. Hunt.

Bermejo-Luque, Lillian (2010) "Second-Order Intersubjectivity: The Dialectical Dimension of Argumentation." *Argumentation.* 24. 85–105.

Bickenbach, Jerome E. and Jaqueline M. Davies (1997) *Good Reasons for Better Arguments.* Orchard Park, NY: Broadview.

Bizer, George Y., Sirel M. Kozak and Leigh Ann Holterman (2009) "The Persuasiveness of the Straw Man Rhetorical Technique." *Social Influence.* 4:3. 216–30.

Black, Katherine (2004) *Idiocy! Taking Conservatives behind the Woodshed.* New York: Authorhouse.

Boghossian, Peter, James Lindsay and Helen Pluckrose (2018) "Academic Grievance Studies and the Corruption of Scholarship." *Aero.* Avalable at: https://areomagazine.com/2018/10/02/academic-grievance-studies-and-the-corruption-of-scholarship/

Bondy, Patrick (2010) "Argumentative Injustice." *Informal Logic.* 30:3. 263–78.

Boudewijn, De Bruin (2013) "Epistemic Virtues in Business." *Journal of Business Ethics,* 113:4. 583–95.

Boudry, Maarten, Fabio Paglieri Massimo Pigliucci (2015) "The Fake, the Flimsy, and the Fallacious: Demarcating Arguments in Real Life." Argumentation. 29. 2. DOI10.1007/s10503-015-9359-1.

Boudry, Maarten (2017) "The Fallacy Fork: Why It's Time to Get Rid of Fallacy Theory." *The Skeptical Inquirer.* 41:5. September/ October, 46–51.

Broad, Jacqueline (2020) "Early Modern Philosophy: A Perverse Thought Experiment." *APA Blog: Women in Philosophy.* October 21. https://blog.apaonline.org/2020/10/21/early-modern-philosophy-a-perverse-thought-experiment/.

Casey, John (2012) "Boethius and the Study of Logic in the Middle Ages." In Philip E. Phillips and Noel Harold Kaylor (eds.), *A Companion to Boethius in the Middle Ages*. Leiden: E.J. Brill, 193–219.
Casey, John (2020a) "Adversariality and Argumentation." *Informal Logic*. 40:1. 77–108.
Casey, John (2020b) "The Persuasive Force of the *Ad Baculum*." *OSSA 12 Conference Archive*. 16. 1–10.
CBS News, 60 Minutes. (2012). Interview with Eric Cantor. http://www.cbsnews.com/8301-18560_162-57348499/the-majority-leader-rep-eric-cantor/?tag=contentMain;cbsCarousel.
Chase, Stuart (1956) *Guides to Straight Thinking*. New York: Harper and Row.
Cohen, Daniel (2001) "Evaluating Arguments and Making Meta-Arguments." *Informal Logic*. 21. 73–84.
Cohen, Daniel (2004) *Arguments and Metaphors in Philosophy*. Lanham: University Press of America.
Cohen, Daniel (2005) "Arguments That Backfire." *The Uses of Argument. OSSA 6 Archives*. 8. 1–10.
Conee, Earl and Richard Feldman (1998) "The Generality Problem for Reliabilism." *Philosophical Studies*. 89:1. 1–29.
Converse, Philip E. (1964) "The Nature of Belief Systems in Mass Publics." In David Apter (ed.), *Ideology and Discontent*. New York: Free Press, 206–61.
Copi, Irving M., Carl Cohen and Daniel E. Flage (2007) *Essentials of Logic, 2e*. Upper Saddle River, NJ: Prentice Hall.
Coulter, Ann (2002) *Slander: Liberal Lies about the American Right*. New York: Three Rivers Press.
Coulter, Ann (2004) *How to Talk to a Liberal (If You Must)*. New York: Crown Forum Books.
Crosswhite, James (1996) *The Rhetoric of Reason: Writing and the Attractions of Argument*. Madison: University of Wisconsin Press.
de Saussure, Louis (2018) "The Straw Man Fallacy as a Prestige-Gaining Device." In Argumentation and Language: Linguistic and Cognitive Explorations. Dordrecht: Springer. 171–90.
Delli Carpini, Michael X. and Scott Keeter (1996) *What Americans Know about Politics and Why It Matters*. New Haven: Yale University Press.
De, Morgan, Augustus (1847) *Formal Logic: Or, The Calculus of Inference, Necessary and Probable*. London: Taylor and Walton.
Dotson, Kristie (2011) "Tracking Epistemic Violence." *Hypatia*. 26. 236–57.
Dretske, Fred (2013) "The Case Against Closure." In M. Steup and J. Turri (eds.), *Contemporary Debates in Epistemology*. 2nd Edition. Malden, MA: Wiley-Blackwell. 27–39.
Dufour, Michel (2019) "Latin Rhetoric and Fallacies." In Bart Garssen, D. Godden, Gordon R. Mitchell and Jean H.M. Wagemans (eds.), Proceedings of the Ninth Conference of the International Society for the Study of Argumentation Sic Sat 2019. Amsterdam: SicSat. 273–83.

Duncombe, M. and C. Dutilh Novaes (2016) "Dialectic and Logic in Aristotle and His Tradition." *History and Philosophy of Logic*. 37:1. 1–8. http://dx.doi.org/10.1080/01445340.2015.1086624.

Ferkany, Matt and Kyle Powys Whyte (2011) "The Importance of Participatory Virtues in the Future of Environmental Education." *Journal of Agricultural and Environmental Ethics*. 25:3. 419–34.

Finocchiaro, Maurice (2005/1981) "Fallacies and the Evaluation of Reasoning." In *Arguments about Arguments*. Cambridge: Cambridge University Press. 109–25.

Finocchiaro, Maurice (2013) *Meta-Argumentation, An Approach to Logic and Argumentation Theory*. London: College Publications.

Flage, Daniel (2004) *Art of Questioning: An Introduction to Critical Thinking*. Upper Saddle River NJ: Pearson.

Fogelin, Robert and Duggan, Timothy J. (1987) "Fallacies." *Argumentation*. 1. 255–62.

Franken, Al (2003) *Lies and the Lying Liars Who Tell Them*. New York: Dutton.

Frankfurt, Harry (2005) *On Bullshit*. Princeton: Princeton University Press.

Fricker, Miranda (2007) *Epistemic Injustice*. New York: Oxford University Press.

Ganapini, Marianna (forthcoming) "The Signaling Function of Sharing Fake Stories." *Mind and Language*.

Gataker, Thomas (1623) A iust defence of certaine passages in a former treatise concerning the nature and vse of lots, against such exceptions and oppositions as have beene made thereunto to Mr. I.B. Wherein the insufficiencie of his answers giuen to the arguments brought in defence of a lusorious lot is manifested; the imbecillitie of his arguments produced against the same further discouered; and the point it selfe in controuersie more fully cleared; London: Printed by Iohn Haviland for Robert Bird, 1623. Ann Arbor: Text Creation Partnership, 2011 https://quod.lib.umich.edu/e/eebo/A01540.0001.001/

Goldberg, Sanford (2018) "Dissent: Ethics and Epistemology." In Casey Rebecca Johnson (ed.), *Voicing Dissent: The Ethics and Epistemology of Making Disagreement Public*. New York: Routledge. 40–60.

Govier, Trudy (1981) "Uncharitable Thoughts about Charity." *Informal Logic*. 4. 3.

Govier, Trudy (1997) *A Practical Study of Argument*. New York: Wadsworth.

Govier, Trudy (1999) *The Philosophy of Argument*. Newport News: Vale Press.

Hacking, Ian (1999) *The Social Construction of What?* Cambridge: Harvard University Press.

Haley, Edward (2006) *Strategies of Dominance: The Misdirection of U.S. Foreign Policy*. Baltimore, MD: Johns Hopkins University Press.

Hamblin, Charles (1970) *Fallacies*. London: Metheuen.

Hansen, Hans V. (2002) "The Straw Thing of Fallacy Theory." *Argumentation*. 16. 133–65.

Hansen, Hans V. and Robert Pinto (1995) *Fallacies: Classical and Contemporary Readings*. State College: Penn State University Press.

Hare, William (2003) "Is It Good to Be Open-minded?" *International Journal of Applied Philosophy*. 17:1. 73–87.

Henning, Tempest (2018) "Bringing Wreck." *Symposion*. 5. 197–211.

Hoven, Paul van den (2015) "Cognitive Semiotics and Argumentation: A Theoretical Exploration." *Argumentation*. 29. 157–76.

Hovhannisyan, Hasmik and Robert Djidjian (2017) "Building a General Theory of Meta-Argumentation." *Metaphilosophy*. 48. 345–54.

Hundleby, Catherine (2009) "Fallacy Forward: Situating Fallacy Theory." In Juho Ritola (ed.), Argument Cultures: Proceedings of OSSA 2009. Windsor, ON: OSSA. 1–10.

Hundleby, Catherine (2010) "The Authority of the Fallacies Approach to Argument Evaluation." Informal Logic. 30:3. 279–308.

Hundleby, Catherine (2013) "Aggression, Politeness, and Abstract Adversaries." *Informal Logic*. 33:2. 238–62.

Hurley, Patrick and Lori Watson (2018) *A Concise Introduction to Logic*. 13th Edition. Boston: Cengage.

Johnson, Ralph and J. Anthony Blair (2006) *Logical Self Defense*. New York: IDEA Press.

Johnson, Ralph and J. Anthony Blair (1983) Logical Self-Defense, 2nd Edition. New York: McGraw Hill.

Johnson, Ralph H. (1987) "In the Blaze of Her Splendors: Suggestions about Revitalizing Fallacy Theory." *Argumentation*. 1. 239–53.

Johnson, Ralph H. (2000) *Manifest Rationality*. Mahwah, NJ: Lawrence Erlbaum Associates.

Kidd, Ian James (2019) "Epistemic Corruption and Social Oppression." In Q. Cassam and H. Battaly (eds.), *Vice Epistemology*. New York: Routledge, 69–85.

Kitcher, Philip (1982) *Abusing Science*. Cambridge: MIT Press.

Krabbe, Erik (2012) "Aristotle's On Sophistical Refutations." Topoi. 31. 243–8.

Krabbe, Erik (2002) "Profiles of Dialogue as a Dialectical Tool." In van Eemeren Frans (ed.), *Advances in Pragma-Dialectics*. Amsterdam: Sic Stat. 153–67.

Laar, Jean-Albert van (2008) "Room for Maneuvering in Raising Critical Doubts." *Philosophy and Rhetoric*. 41:3. 195–211.

LaMarre, Heather L., Kristen D. Landreville and Michael A. Bream (2009) "The Irony of Satire: Political Ideology and Motivation to See What You Want to See in 'The Colbert Report'." *The International Journal of Press/Politics*. 14. 212–31.

Lambert-Beatty, Carrie (2009) "Make-Believe: Parafiction and Plausibility." October. 129. 51–84.

Leibowitz, Uri (2013) "Particularism in Aristotle's *Nicomachean Ethics*." *The Journal of Moral Philosophy*. 10:2. 121–47.

Lewiński, Marcin (2014) "Argumentative Polylogues: Beyond Dialectical under- Papers Standing of Fallacies." *Studies in Logic, Grammar and Rhetoric*, 36:1. 193–218.

Lewiński, Marcin (2011) "Towards a Critique-Friendly Approach to the Straw Man Fallacy." *Argumentation*. 25. 469–97.

Lewiński, Marcin (2020) "The Straw Man and its Baby Semantics." In J. Anthony Blair and Christopher W. Tindale (eds.), *Rigour and Reason: Essays in Honour of Hans V. Hansen*. Windsor: Windsor Studies in Argumentation. 276–303.

Lewiński, Marcin and Steve Oswald (2013) "When and How Do We Deal with Straw Men? A Normative and Cognitive Pragmatic Account." *The Journal of Pragmatics*. 59. 164–77.

Locke, John (1975) *An Essay Concerning Human Understanding*. Edited by P.H. Nidditch. Oxford: Oxford University Press.

Lord, Jeffrey (2007) "Iraq and the Party of Race." In *The American Spectator*. February 16.

Massey, Gerald (1981) "The Fallacy behind Fallacies." *Midwest Studies in Philosophy*. 6. 489–500.

Medina, Jose (2013) The Epistemology of Resistance. New York: Oxford University Press.

Mill, John Stuart (1882) *A System of Logic. Ratiocinative and Inductive, being a Connected View of the Principles of Evidence, and the Methods of Scientific Investigation*. 8th Edition. New York: Harper & Brothers, Publishers, Franklin Square.

Mill, John Stuart (1991) *On Liberty*. Indianapolis: Hackett.

Moore, Brooke Noel and Richard Parker (2004) *Critical Thinking*. 7/e. New York: McGraw Hill.

Nagel, Jennifer (2019) "Epistemic Territory." *Proceedings and Addresses of the American Philosophical Association*. 93. 67–83.

Nolt, John (1997) *Logics*. Belmont, CA: Wadsworth.

Oswald, Steve and Marcin Lewiński (2014) "Pragmatics, Cognitive Heuristics, and the Straw Man Fallacy." In Thierry Herman and Steve Oswald (eds.), Rhetoric and Cognition. Bern: Peter Lang. 313–43.

Paglieri, Fabio (2013) "Choosing to Argue." *The Journal of Pragmatics*. 59. 153–63.

Paul, Richard (1984) "Teaching Critical Thinking in the 'Strong' Sense." *Informal Logic Newsletter*. 4:2. 2–7.

Peirce, Charles Sanders (1877) "The Fixation of Belief." *Popular Science Monthly*. 12. 1–15. Cited from *Collected Papers Volume 5*. Edited by Charles Hartshorne and Paul Weiss. Cambridge: Harvard University Press. Cited in-text in accord with standard methods of reference to Peirce's edition as: CP Vol#.Page#.

Perelman, Chaim (1982) The Realm of Rhetoric. Trans. W. Kluback. Notre Dame: Notre Dame University Press.

Perelman, Chaim and Lucy Olbrechts-Tyteca (1969) *The New Rhetoric: A Treatise on Argumentation*. Trans. J. Wilkinson and P. Weaver. Notre Dame: University of Notre Dame Press.

Peter of Spain (2014) "*Summaries of Logic*." In Brian Copenhaver, Calvin Normore and Terence Parsons (eds.), *Peter of Spain: Summaries of Logic: Text Translation, Introduction and Notes*. Oxford: Oxford University Press, 100–509.

Prichard, Harold A. (1912) "Does Moral Philosophy Rest on a Mistake?" *Mind*. 21:81. 21–37.

Ribeiro, Brian (2008) "How Often Do We (Philosophy Professors) Commit the Straw Man Fallacy?" *Teaching Philosophy*. 31. 27–38.

Rooney, Phyllis (2010) "Philosophy, Adversarial Argumentation, and Embattled Reason." *Informal Logic*. 30:3. 203–34.

Rooney, Phyllis (2012) "When Philosophical Argumentation Impedes Social and Political Progress." *Journal of Social Philosophy*. 43:3. 317–33.

Rudinow, Joel and Vincent E. Barry (2008) *Invitation to Critical Thinking*. 6e. Belmont, CA: Wadsworth.

Savage, Michael (2005) *Liberalism Is a Mental Disorder*. New York: Nelson Current.

Schechter, Joshua (2019) "Small Steps and Great Leaps in Thought: The Epistemology of Basic Deductive Rules." In Magdalena Balcerak Jackson, Brendan Balcerak Jackson (eds.), *Reasoning: New Essays on Theoretical and Practical Thinking*. Oxford: Oxford University Press, 152–75. DOI:10.1093/oso/9780198791478.003.0009.

Schumann, Jennifer and Sandrine Zufferey (2020) "Connectives and the Straw Man: Experimental Approaches in French and English." *OSSA 12 Archives*, 1–20.

Schumann, Jennifer, Sandrine Zuffrey, and Steve Oswald (forthcoming) "The Linguistic Formulation of Fallacies Matters." *Argumentation*.

Schumann, Jennifer, Sandrine Zuffery and Steven Oswald (2020) "Connectives and Straw Men: Experimental Approaches in French and English." In *OSSA Archives*. Windsor: University of Windsor, 6.

Scott, Kyle (2014) "The Political Value of Humility." *Acta Politica*. 49. 217–33.

Scriven, Michael (1976) *Reasoning*. New York: McGraw Hill.

Siegel, Harvey (1995) "What Price Inclusion?" *Teacher's College Record*. 97. 6–31.

Sinnott-Armstrong, Walter (2004) "Classy Pyrrhonism." In Walter Sinnot-Armstrong (ed.), *Pyrrhonian Skepticism*. Oxford: Oxford University Press. 188–207.

Sinnott-Armstrong, Walter (2008) "A Contrastivist Manifesto." *Social Epistemology*. 22:3. 257–70.

Snedegar, Justin (2013) "Reason Claims and Contrastivism about Reasons." *Philosophical Studies*. 166. 231–42.

Snedegar, Justin (2015) "Contrastivism about Reasons and Ought." *Philosophy Compass*. 10. 379–88.

Sokal, Alan and Jean Bricmont (1997) *Fashionable Nonsense*. New York: Picador.

Somin, Ilya (1998) "Voter Ignorance and the Democratic Ideal." *Critical Review*. 12:4. 413–58.

Spear, Andrew (2020) "Gaslighting, Confabulation, and Epistemic Innocence." *Topoi*. 39. 229–41.

Stark, Cynthia (2019) "Gaslighting, Misogyny, and Psychological Oppression." *The Monist*. 102. 221–35.

Stevens, Katharina (2020) "Principle of Charity as a Moral Requirement." In *OSSA Archives*. Windsor: University of Windsor, 19.

Stevens, Katharina (2021) "Fooling the Victim: Of Straw Men and Those Who Fall for Them." *Philosophy & Rhetoric*. 54:2. 109–27.

Sunstein, Cass (2009) *On Rumors*. New York: Farrar, Straus, and Giroux.

Tanesini, Alessandra (2018) "Eloquent Silences: Silence and Dissent." In Casey Rebecca Johnson (ed.), *Voicing Dissent: The Ethics and Epistemology of Making Disagreement Public*. New York: Routledge. 109–27.

Teays, Wanda (2006) *Second Thoughts: Critical Thinking for a Diverse Society*. New York: Mc Graw Hill.

Tindale, Christopher (1999) *Acts of Arguing: A Rhetorical Model of Argument*. Albany: SUNY Press.

Tindale, Christopher (2004) *Rhetorical Argumentation*. London: Sage Publishing.

Tindale, Christopher (2007) *Fallacies and Argument Appraisal*. Cambridge: Cambridge University Press.

Tindale, Christopher (2015) *The Philosophy of Argument and Audience Reception*. Cambridge: Cambridge University Press.

van Eemeren, Frans and Rob Grootendorst (1987) "Fallacies in a Pragma-Dialectical Perspective." *Argumentation*. 1. 283–301.

van Eemeren, Frans and Rob Grootendorst (2004) A Systematic Theory of Argumentation. Cambridge: Cambridge University Press.

van Eemeren, Frans and Rob Grootendorst and Francisca Snoeck Henkenmans (2002) *Argumentation: Analysis, Evaluation, Presentation*. Mahwah, NJ: Lawrence Erlbaum.

Vaughn, Lewis (2008) *The Power of Critical Thinking, 2e*. Oxford: Oxford University Press.

Vernon, Thomas S. and Lowell A. Nissen (1968) *Reflective Thinking*. Belmont: Wadsworth.

Walton, Douglas (1989a) "Dialogue Theory for Critical Thinking." *Argumentation*. 3:2. 169–84. DOI:10.1007/BF00128147.

Walton, Douglas (1989b) *Informal Logic*. Cambridge: Cambridge University Press.

Walton, Douglas (1992) *Plausible Argument in Everyday Conversation*. Albany: SUNY Press.

Walton, Douglas (1995) *A Pragmatic Theory of Fallacy*. Tuscaloosa: Alabama University Press.

Walton, Douglas (1996) "The Straw Man Fallacy." In van Bentham Johan, van Eemeren Frans, Rob Grootendorst and Frank Veltman (eds.), *Logic and Argumentation*. Amsterdam: Royal Netherlands Academy of Arts and Sciences, North Holland. 115–28.

Walton, Douglas (1998) *Ad Hominem Arguments*. Tuscaloosa: University of Alabama Press.

Walton, Douglas (1999) *Informal Logic: A Handbook for Critical Argumentation*. Cambridge: Cambridge University Press.

Walton, Douglas (2013) *Methods of Argumentation*. Cambridge: Cambridge University Press.

Walton, Douglas, Chris Reed and Fabrizio Macagno (2008) *Argumentation Schemes*. Cambridge: Cambridge University Press.

Walton, Douglas and Fabrizio Macagno (2010) "Wrenching from Context: The Manipulation of Commitments." *Argumentation*. 24. 283–317.

Watts, Isaac (1824) *Logic; or, The Right Use of Reason*. London: J. Rivington.

Weinstock, Michael, Yair Neuman and Iris Tabak (2004) "Missing the Point or Missing the Norms? Epistemological Norms as Predictors of Students' Ability to Identify Fallacious Arguments." Contemporary Educational Psychology. 29. 77–94.

Whately, Richard (1826) *Elements of logic*. London: B. Fellowes.

Wittgenstein, Ludwig (1953) *Philosophical Investigations*. Trans. G.E.M. Anscombe. Englewood Cliffs: Prentice Hall.

https://georgewbush-whitehouse.archives.gov/stateoftheunion/2004/

https://georgewbush-whitehouse.archives.gov/news/releases/2004/04/20040430-2.html

Index

A
Ad baculum (fallacy) 53, 85, 127, 161
Ad hominem (fallacy) 30, 47, 48, 53, 56, 60, 76, 78, 79, 85, 87, 106, 114, 127, 137, 161, 162–3, 209–10
  Tu quoque (fallacy) 205
Ad verecundiam (fallacy) 30, 32, 47, 87, 161
Abramson, Kate 131–3
adversariality (of argument) 6, 21–8, 31–3, 36, 38, 44, 76, 86, 136, 190
  contrastive 25, 142–3, 200
  Doxastic 5–6, 21–7, 200–1
  minimal dialectical 5, 21–2, 25–7, 120, 200–1
agreement, problem of 11
Akerman, Bruce 68
Anna Karenina Problem, the 11, 17, 23, 33, 40, 55–6, 106
Aristotle 8, 30, 33–8, 40, 44, 49, 50, 56, 85, 109, 163, 194, 199. 201, 213
Arnauld, Antoine 37–9
Aspetitia, Barceló 205
asserting the consequent (fallacy) 1
Astell, Mary 40
audience, onlooking 42–6, 49, 62–3, 69, 76–7, 101, 103, 105, 108, 113–19, 123, 125, 126, 130–1, 137, 143, 149, 158–9, 168, 181, 185, 192, 195–9, 213
Avenatti, Michael 170

B
Bach, J. S. 11
backfire problem (for Iron Man) 99, 105
Balmford, J. 40
Baronett, Stan 46
Barrett, Amy Coney 206–9
Battaly, Heather 111
Beardsley, M. C. 45–6
Biden, Joseph 124–5
Bizer, George 77, 183, 197–8

Black, Lewis 69
Blair, J. Anthony 46, 48, 183
Boghossian, Peter 172–5
Bondy, Patrick 16
Bothsiderism 12, 19, 179, 203, 204–9, 214
Boudry, Martin 8, 10
Broad, Jacqueline 40
Brzezinski, Mika 66–7
Bullshit 179
Bush, George H.W. 13, 73

C
Cantor, Eric 95–6
Carruthers, Charlene. 99
Carson, Ben 156–7
charity,
  argumentative 4, 61, 64, 145, 174, 190
  toxic 50, 86, 98, 102–8, 114
Chase, S. 57
clearing the decks 91–3, 152, 175, 179, 185
Clinton, William J. 13
closing function (of straw man arguments) 59, 84, 90, 92, 94, 96, 113, 146, 151–2, 155–60, 168, 179, 186, 192, 208, 213
Cohen, Carl 46
Cohen, Daniel 99, 127
Colbert, Steven 171
contrastivism 25, 50, 140–3, 154–6, 176, 182, 200, 205–6, 213
convergent arguments 57, 59–60
Copi, Irving 46, 51
Coulter, Anne 69, 75–8, 162–3, 209
Crosswhite, James 195

D
Daniels, Stormy 169–70
Deliberation day 68
de Beauvoir, Simone 132
DeMorgan, Augustus 41, 44–6
dialecticality, puzzle of 2, 49, 112–14, 139–52, 191, 213

diving (gamesmanship) 19, 168–70, 176–80, 201
Dixon, Benjamin 118
doctrine of compensation (Aristotle) 109–10
Durden, Tyler 67
Dyadism (of dialogue) 40, 42, 101, 126, 136, 143, 181–94, 203, 213

E
Eager Beaver Problem 21, 201, 203, 214
effectiveness, puzzle of 1–2, 36, 42, 45, 47–9, 56, 61, 77–8, 94, 113–37, 139, 158, 174, 181, 184–5, 189, 191–9, 201, 213
Epicurus 192
Epistemetric considerations 99, 103, 105–6, 127, 185

F
fallacy fallacy 18–20, 203
fallacy fallacy fallacy 19–20, 210
fallacy fork 8, 10, 13, 15
fallacy paradox 13–15, 203, 209
Finocchiaro, Maurice 8, 9, 12, 202
fishing (conversational) 83
Fishkin, James 68
Flage, Daniel 46
Franken, Al 69
free speech fallacy 19, 30–1, 179
Fricker, Miranda 16
Fukuyama, Francis 64

G
gaslighting 78, 114, 121, 130–7, 185, 193
Gataker, Thomas 40
generality problem 5, 7–15, 16, 27–8
Goldberg, Jonah 165
Goldberg, Sanford 129
Govier, Trudy 24–5, 46, 48, 102, 200
Groarke, Leo 46
Grootendorst, Rob 6, 56, 189, 191

H
Haley, Edward 74
Hamblin, Charles 6, 30, 35
Hannity, Sean 65, 148–9
Hanson, Victor Davis 163
Hare, William 109

Harris, Kamala 206–7
Harry Potter Problem 18, 30–4, 161, 201, 203, 209, 211, 214
hasty generalization (fallacy) 1, 2, 31, 57–60, 69, 143, 156
Hegel, G.W.F. 210
Heer, Jeet 177
Henning, Tempest 16
hermeneutics of antipathy 11, 26, 43, 46, 67, 77, 119–20, 123, 125–6, 128, 136, 137, 146–8, 152, 158, 159, 170
Hoft, Jim 95
Holder, Eric 65
Holterman, Leigh Ann 197
Huckabee, Michael 164
Hundleby, Catherine 16, 20
Hurley, Patrick 47

I
Ignoratio elenchi 17, 30, 34–41, 44–5, 49, 51, 56, 85, 199, 213
Iron man 4, 12, 18, 29, 85–99, 101–11, 113, 114, 146, 153, 170, 181, 186, 201, 206, 208, 213

J
Jack Straw 40
Johnson, Ralph 6, 8, 15, 46, 48, 183

K
Kidd, Ian James 111
Kozak, Sirel 77, 183, 197–9

L
Lewiński, Marcin 64, 80, 183, 190–2
Lewis, Helen 177–8
Lord, Jeffrey 74

M
Macagno, Fabrizio 87, 106, 118
Massey, Gerald 8
McConnell, Mitch 124
Meletus 147
meta-argumentation 19–21, 34, 41, 43–4, 56, 58–9, 66, 70, 72, 78, 84, 87, 92, 96, 104, 112–13, 139–40, 143, 146, 152–80, 181–2, 191–4, 200–3, 205, 208, 211, 213–14

strong sense 161–8, 174, 180–2, 209–10, 209, 214
meta-argumentation, puzzle of 1–2, 49
metaphilosophy 19, 210–12
Mill, John Stuart 41, 44–5, 63, 163, 208
Modus tonens 45, 55
Moore, B.N. 47
moral hazard problem (for Iron Man arguments) 4, 98–9, 105, 108
mutuality thesis 21–4, 27

N
Nagel, Jennifer 83
negativity problem 5, 20–8
Neumayr, George 121
Nicole, Pierre 37–9, 42, 46, 47, 49, 55
Nissen, L.A. 57
Noah, Trevor 66

O
Obama, Barack 13, 64–5, 121–5, 163–5, 209
Obama, Michelle 65
Olbrechts-Tyteca, Lucie 195
open-mindedness 106–11
Oswald, Steve 64, 183, 190–2, 198, 199
Owl of Minerva Problem 18, 19, 32–5, 209–14

P
Paglieri, Fabio 8, 9, 15, 127
Palin, Sarah 171
Parker, R. 47
Peirce, C.S. 23
Peirce Problem 23–4
Perelman, Ch. 195
personal address (of straw man)
   forms 114–15, 139
   mixed cases 123–5
   second person 115, 120–37, 143, 160, 172–3, 185
   third person 115, 120, 126, 143, 158–9, 181, 185
Pigliucci, Massimo 8, 9, 15
Plato 20, 36, 89, 147
Poe's Law 31, 171–2
polarization 77–8, 159
Polyadism (of dialogue) 40, 46, 101, 126, 136, 143, 181, 185, 188, 195, 213
Prager, Dennis 158–9

Pragma-dialectics 6, 183, 189–92, 203, 213

R
Reagan, Ronald 95
Ribeiro, Brian 12, 82, 88
Rooney, Phyllis 16, 20–1, 24–5

S
Sartre, Jean Paul 132, 177
de Saussure, Louis 183, 193–4
Savage, Daniel 69
Schumann, Jennifer 183, 198–9
scope problem 5, 15–20, 23, 27, 34, 45
scorekeeping, argumentative 4, 39, 96–7, 157
Scriven, Michael 46
silencing 4, 48, 78, 114–15, 126–30, 137, 139, 143, 160, 185, 193, 199
Slayer (hail Satan!) 11
slippery slope (fallacy) 27, 91
Socrates 86, 147–8
Sokal, Alan 171–4
Stahl, Leslie 95–6, 104
standpoint rule 189–90
Stark, Cynthia 133
Straw man
   burning 3, 29, 40, 114, 117, 147–52, 186, 203, 212
   hollow 3, 14, 29, 42, 47, 50, 53, 70–9, 84–5, 91–3, 107, 113, 136, 157, 159, 165, 181, 186, 188–9, 212
   selectional/weak 3, 14, 29, 42, 50, 53–4, 58–70, 77, 79, 82, 181, 186–8, 208, 212
   self 3, 169–71, 175–80
Sullivan, Andrew 64
Sunstein, Cass 78, 79
Superaddressee 195–6
swamping (argumentative) 149

T
Talisse, Robert B. 19, 33, 43, 45, 55, 149
Tesfaye, Sophia 156–7
Tindale, Christopher 46, 79, 87–8, 183, 195–6
Tolstoy, Leo 33, 34
Trump, Donald J. 12–13, 63–4, 66–7, 76, 115–19, 124–5, 145–6, 169–71, 176, 178, 204, 206

V

Van Eemeren, Frans 6, 56, 183, 189, 191
Van Laar, Jean Albert 80
Van Norden, Bryan 150–2
Vernon, T.S. 57

W

Walton, Douglas 6, 9, 57, 82, 84, 87, 106, 118, 183–9, 203
Watson, Paul Joseph 145
Watters, Jessie 170
Watts, Isaac 30, 39–40, 45, 86
weaponization (of metalanguage) 32–3, 176
West, Allen 204
Westboro Baptist Church 96–7
Wexler, Chuck 99
Whataboutism 12, 179, 204–6
Whately, Richard 13, 41
Wilhelm, Heather 63
Woodruff, Judy 99
Woolf, Michelle 145

Z

Zuffrey, Sandrine 183

www.ingramcontent.com/pod-product-compliance
Lightning Source LLC
Chambersburg PA
CBHW062214300426
44115CB00012BA/2061